Production Processes

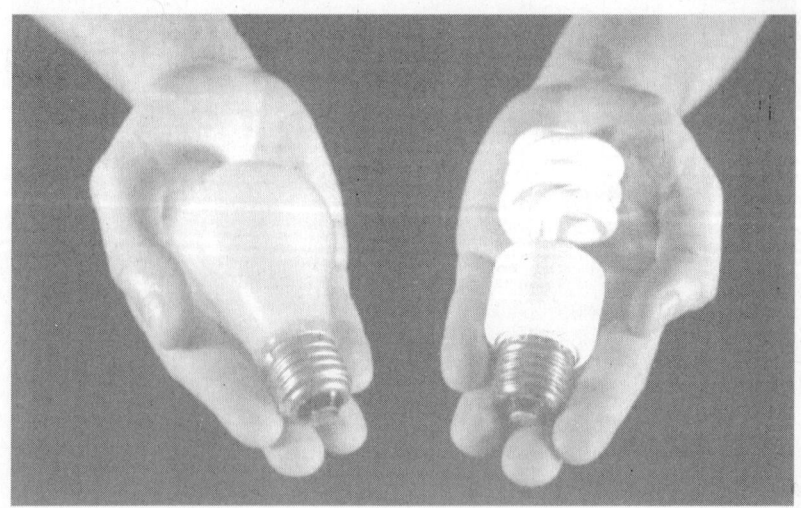

Production Processes

Midhat Luqman

Life Fellow
Institution of Engineers (India)
Allahabad

CBS

CBS Publishers & Distributors Pvt. Ltd.

New Delhi • Bengaluru • Chennai • Kochi • Kolkata • Mumbai
Hyderabad • Uttarakhand • Nagpur • Patna • Pune • Jharkhand

ISBN: 978-81-239-1898-3

First Edition: 2010
Reprint: 2019

Published by **Satish Kumar Jain** and produced by **Varun Jain** for **CBS Publishers & Distributors Pvt. Ltd.,**
4819/XI Prahlad Street, 24 Ansari Road, Daryaganj, New Delhi - 110002
delhi@cbspd.com, cbspubs@airtelmail.in • www.cbspd.com
Ph.: 23289259, 23266861, 23266867 • Fax: 011-23243014

Corporate Office: 204 FIE, Industrial Area, Patparganj, Delhi - 110 092
Ph: 49344934 • Fax: 011-49344935
E-mail: publishing@cbspd.com • publicity@cbspd.com

Branches:
• *Bengaluru:* 2975, 17th Cross, K.R. Road, Bansankari 2nd Stage,
  Bengaluru - 70 • Ph: +91-80-26771678/79 • Fax: +91-80-26771680
  E-mail: cbsbng@gmail.com, bangalore@cbspd.com
• *Chennai:* No. 7, Subbaraya Street, Shenoy Nagar, Chennai - 600030
  Ph: +91-44-26681266, 26680620 • Fax: +91-44-42032115
  E-mail: chennai@cbspd.com
• *Kochi:* Ashana House, 39/1904, A.M. Thomas Road, Valanjambalam,
  Ernakulum, Kochi • Ph: +91-484-4059061-65
  Fax: +91-484-4059065 • E-mail: cochin@cbspd.com
• *Kolkata:* 6-B, Ground Floor, Rameshwar Shaw Road, Kolkata - 700014
  Ph: +91-33-22891126/7/8 • E-mail: kolkata@cbspd.com
• *Mumbai:* 83-C, Dr. E. Moses Road, Worli, Mumbai - 400018
  Ph: +91-9833017933, 022-24902340/41 • E-mail: mumbai@cbspd.com

Representatives:

• Bhubaneswar 0-9911037372	• Hyderabad 0-9885175004	• Jharkhand 0-9811541605
• Nagpur 0-9021734563	• Patna 0-9334159340	• Pune 0-9623451994
• Uttarakhand 0-9716462459	• Dhaka (Bangladesh) 01912-003485	

Printed at:
India Binding House, Noida, UP (India)

Preface

In the last few years the country has made tremendous progress by not leaving any recognisable sphere of development unattended. This has led to immense national prosperity and we have moved a step closer to attain the status of a developed country. This has only been possible because existing technologies and new 'know-how' have worked well together, thereby contributing to 'sustainable development'. The best use of technology is possible when sufficient knowledge and know-how is made available and is utilised well by technocrats.

Books have always been the proven source of knowledge. There are a great number of books already on the market which give the readers excellent information on theoretical aspects of technologies and production processes. But theoretical knowledge unless one knows how it is to be implemented in practice, is incomplete knowledge. This book has, therefore, been designed to fulfil this need and, as such, it throws light on various production processes on shopfloor. It provides insight into practical handling of various processes on floor and consequently can be expected to serve as an ideal handbook beneficial to all floor-workers, managers as well as new generation of technocrats.

As far as production processes are concerned, the sky is the limit and it is very difficult for a single author to write the minute details of each and every process. In this book a sincere effort has been made to look into the production processes and their allied activities namely conversion of plastic raw material into product, press working of sheet metal, impact extrusion of aluminium and zinc, vacuum metalising of reflectors, miniature lamp making technology and process, quality assurance in production processes and automation, etc.

This book, should serve as an encouragement to other shopfloor engineers to translate their experiences into similar books for the benefit of students, entrepreneurs and industrialists.

It is advisable that for insight into technical and commercial information, specifications and properties, data may be searched on internet. Details of process, plant, machineries and equipment may vary from shopfloor to shopfloor. It is, therefore, necessary that readers of this book should verify information from relevant manuals which are generally provided by the process providers, plant and machinery makers and suppliers.

Midhat Luqman

Acknowledgements

Writing a book for publication needs good planning and fixing a target time frame for submission of work to the publisher. At the planning stage it was realized that to prepare book 'Production Processes' author would need support from his well-wishers.

It would be discourteous on my part if do not mention organizations and my friends who whole-heartedly supported me in preparing the manuscript. First of all I would like to thank Mr Kaleem Ahmad Khan, General Manager who on behalf of M/s Shervani Industrial Syndicate Limited (SISL) allowed me to take a few snaps of machines, tools and automation for this book. My thanks are also due to M/s Schuler AG Germany for providing few photographs of Impact Extrusion Press and System.

Mr Waseem Ahmad Khan, my friend and ex-colleague, helped me in tieing up with the publisher. My ex-colleague Mr Suhail Ansari helped me untiringly in producing drawings by Auto Cad.

My thanks are also due to Prof K.K. Bhutani, Director, UPTECH for giving me valuable suggestions in planning the book and preparing the introduction. I have all the praises and best wishes for the daughter-in-law of my close friend Mr Bipin Tandan, Mrs Anju Rahul Tandan, Principal Army School, Allahabad. She helped me in giving finishing touches to the manuscript and preface.

In the last but not the least, I want to thank my wife Nuzhat Luqman who supported and encouraged me during this ordeal, especially in making the monogram effective and impressive. My children also remained a symbol strength for me during the entire course.

Midhat Luqman

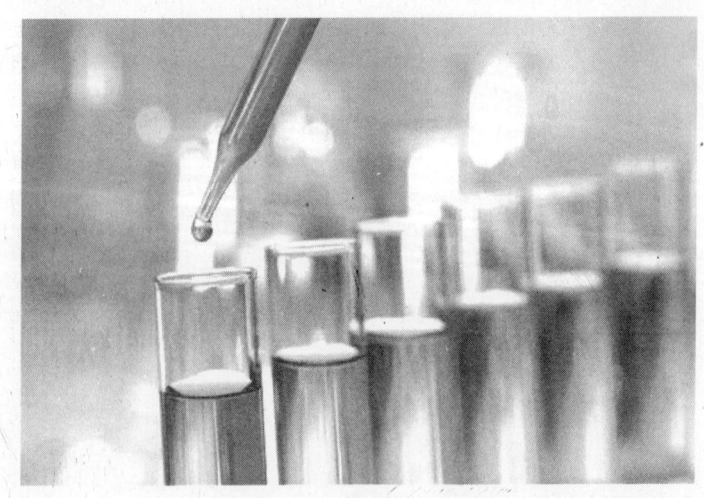

Contents

List of Figures

Introduction

It is rightly said that seeing leads to believing and practice leads to perfection. On any shopfloor, practical wisdom is derived by repeated operations. We have a large number of engineers who have contributed in the advancement of technology but there are a few who have translated their experiences in the form of books and monographs. Production is directly linked with prosperity and production processes are the ways and means of achieving prosperity, both for an industry and for a nation.

Theoretical knowledge is useful for understanding the processes but practical knowledge opens up reality. Abundant literature is available to explain and elaborate the laws of mechanics involved in producing an object but there is still a need to explore and explain the most suitable processes the engineer must adopt in producing quality products at reasonable cost. Study of various production processes is necessary and useful to pinpoint suitability in shorter period of time. In the present book the author has shared his experiences of shopfloor in order to provide useful and beneficial know-how to the practising engineers and technocrats at large.

Production processes are innumerable having different parameters. It is not easy and may not ever be correct to pinpoint the best one. With the change of coordinates of space and time, the selective processes may require different logical decisions. In present work the author has undertaken a few of the important processes to cover the wider canvas of shopfloor of different industries. It is hoped that the writing will go a long way to benefit students, entrepreneurs and the industry in making a right choice
of shopfloor facilities.

Conversion of Plastic Raw Materials into Product

Plastics and their Development

Plastics raw materials are not generally seen in the market. There are specialized wholesale and retailing business houses and shops where plastic raw materials of various types and grades are available. These are in the shape of granules, resins and powder. Granules are approximately round-shaped balls. Dimensions of granules vary from say, one millimetre to three millimetres. Granules may also be cylindrical in shape, cut from round strands. Diameter of such granules vary approximately from two millimetres to three millimetres. Plastic raw material in the form of small particles is termed as resins, in the form of minute particles, as powder. Shapes, dimensions, colour of granules depend on types and grades of plastics.

Development of plastics has a history which dates back to the days of Babilonian era. Shellac, a natural material, like plastics, was used to make ornaments like bangles. Shellac becomes soft when heated and hard again when allowed to cool down. Gutta percha, a plastic, was developed around the beginning of the nineteenth century. It was produced by chemically reacting cellulose with chemicals. Around this time it was realized by the scientists and technologists that plastics have a great potential to become a raw material for producing a variety of intermediate and end products. These offered great flexibility in creating shapes, sizes, quantities and aesthetics of products which normally is very difficult to achieve if produced in metals and other materials. Moreover, conversion of plastic raw materials to products or components is much less expensive as compared to producing the same or similar parts in metal. With the realization of these advantages, great emphasis was given to development of a variety of plastic raw materials and their production during middle of nineteenth century. Around the same time exploration and development work was going on for petroleum by products. Combination of both the efforts gave rise to development and production of a variety of plastic raw materials from petroleum byproducts. Crude is extracted from the depths of earth by boring special borewells and sea beds. Extracted petroleum is in the form of viscous blackish sludge. It is then transferred to nearby refineries for refining it to obtain various byproducts like naptha, crude oil, gases and other liquids. Plastics which were initially developed were polyvinyl chloride, polyethylene, polystyrene. The emphasis now has shifted to development of various qualities in these plastic materials. With passing time, plastic processing industries became organized and demanded 'tailormade' plastic raw materials with specific quality norms. This gave rise to a number of varieties and grades of plastics.

Types of Plastics

There are two families of plastics. One is thermosetting and other is thermoplastics. Plastics of thermosetting category can be heated once to form a shape when pressed in a heated mould for a particular period of time. Reheating of thermosetting will not soften it. If heating is continued for a longer period of time, thermosetting plastics would degrade.

Thermoplastics are those plastics which may be heated many times to get soften. Degradation in quality of thermoplastics does take place if softening temperature is not precisely maintained, if exposure to heat is for much longer period of time. Heated thermoplastics can also be degraded due to contamination by moisture, dust, fumes, odor, etc.

Commonly used thermosetting plastics are urea formaldehyde, amino formaldehyde, melamine, etc. Components produced from thermosetting plastics are hard and flame-resistant, generally used for producing electrical switches, contractor housing and other components requiring hardness and resistant to burning. In thermoplastics category, there are a large number of varieties and grades. Some of them are extrusion grades, moulding grades, flame retardants, medium impact, high impact, plating grade, glass fibre reinforced, EMI shielding (conductive), homopolymers, copolymers, mineral-filled copolymers, PTFE-filled homo and copolymers, UV-stabilized homopolymers, heat-resistant, super tough, low viscosity, high viscosity, graphite fibre-reinforced, etc. Indian and international plastics-producing companies of repute produce a large number of grades having variations in properties. Injection moulders, desirous of getting detailed information may refer to magazines, catalogues, books on thermoplastics and thermosetting plastics. Sufficient information can also be obtained from reputed producers of plastics in India, Japan, Germany, UK, USA, Italy, France and many other countries. Names and addresses of international plastics producers may be searched on various search engines like Google, Yahoo on Internet. Various properties parameters of resins and compounds may be studied from literature on plastic material properties. On study of above it may be appreciated that a large number of grades can be produced by varying various properties parameters.

Some plastic materials are of general purpose and others are meant for highly specialized purposes. Some commercially available thermoplastics are polyethylene, polypropylene, polystyrene, acrylonitrile butadine styrene, polycarbonate, acrylic, polyamides, polyethylene terephthalate, PVC and a large number of modified grades which are produced by blending various thermoplastics, additives, modifiers, plasticizers, fillers, pigments, flame retardant agents. Highly specialized types of plastics are those which are used for medical care equipment, human body spare parts, outer space vehicles and satellite parts and instruments used in severely cold environment, high temperature environment usage. A few specialized thermoplastics are acrylic polymers having low level of allergenicity, PP, PUR, EAM, etc.

Properties of Plastics

Properties of plastics may be defined under following headings:

   Mechanical
   Electrical
   Environmental
   Optical

Mechanical: Tensile strength
                   Compressive strength
                   Flexural strength
                   Hardness
                   Specific gravity
                   Melt flow index
                   Impact strength
                   Toughness
                   Creep strength

Electrical: Electrical conductivity
                   Electrical insulation
                   Dielectric constant

Environmental: Effect due to exposure to atmosphere
                   Effect of ultraviolet and infrared energy
                   Behaviour in vacuum and weightlessness in
                   outer space
                   Effect of sea water pressure
                   Effect of microorganism
                   Biodegradation

Optical: Refractive index
                   Transparency
                   Colourability
                   Effect of visible light

Tests for almost all the properties are standardized in India and many other countries. Bureau of Indian Standards has standardized test methods, testing equipments and ranges of various properties. Most of the plastic raw material producers specify range of various properties of different types and grades of plastics. Generally, producers of plastics provide technical brochures for prospective buyers. Buyers can study properties of plastics from various producers and select most suitable type and grade.

Varieties and Grades of Plastics

Polyethylene generally has three varieties, namely low density, linear low density and high density. There are three important variables on which quality depends. It is specific gravity, melt flow index and processing temperatures. Polyethylene can be converted into useable product by processes like injection moulding, blow moulding, extrusion, thermoforming, rotational moulding, etc.

High density polyethylene has a specific gravity around 0.96. It is tough and rigid as compared to low density polyethylene which has specific gravity around 0.93. High density polyethylene retains its rigidness to a great extent even at a temperature of 100 degree centigrade. Like low density polyethylene, HDPE also has a wax-like feeling if touched and can be scratched. HDPE producing companies generally produce a number of grades having variation in properties to cater to the needs of processors. Some grades of HDPE are suitable for injection moulding process, some for extrusion and some for blow moulding. In injection moulding variety of HDPE, there are a number of grades having variation in property norms. For example, for moulding thin-walled components, a grade with higher melt flow index would suit.

Polypropylene has a specific gravity around 0.91 to 0.92. Hence more numbers of components can be moulded as compared to HDPE. Polypropylene (PP) is tough and hard as compared to HDPE. Consequently, it cannot be scratched easily. It is also less flexible as compared to HDPE.

Polystyrene is normally a clear, hard and brittle plastic having a specific gravity around 1.12. It is known as general purpose polystyrene. By adding plasticizers, rubber and modifiers, a high impact grade is produced. It is called high impact polystyrene. With this grade of polystyrene shiny lustrous surface cannot be achieved in injection moulded components and articles.

Acrylonitrile butadiene styrene (ABS) is considered as engineering plastic by virtue of its properties. It is tough, hard and give lustrous surface when moulded in a highly polished cavity of injection mould. ABS is available in a number of grades to meet various requirements of processors, namely for extrusion, thermoforming and plating, etc. Normally ABS is coloured during production process to give attractive shades standardized by producer of ABS.

Polymethyl methacrylate (acrylic) is a rigid and somewhat tough clear plastic. Its light transmittance is around 98%. Its specific gravity is 1.24 and melt flow index ranges from 1.4 to 27. A number of transparent colour shades can be obtained by mixing master batches at mixer-loader of automatic screw-plunger injection moulding machines. Depending upon condition of cavities surfaces of injection moulds, Acrylic (PMMA) provides good surface finish of moulded articles. This is also termed as engineering plastics.

Polycarbonate is somewhat rigid and tough clear plastic. Its light transmittance is around 97%. Specific gravity is 1.2 and melt flow index ranges from 3 to 10 g/10 min. A number of transparent colour shades can be obtained by mixing master batches at mixer-loader of automatic screw-plunger injection moulding machine. Fine filaments can also be extruded from extrusion grades of polycarbonate. It is a clear and tough plastic. It has some advantage in properties over acrylics . It is considered an engineering plastic by virtue of its properties like good mouldability, low percentage of shrinkage, ambient heat and light stability, good colourability, less migration of pigments. It is also a good plastic for extrusion process.

Polyamides (nylon) are many injection moulding grades including glass fibre filled grades. Polyamide 6, 66, 6 12 and 46 are a few grades with different properties. Polyamide is very hygroscopic plastic therefore, it should be effectively dried before injection moulding of a variety of components. Polyamides offer much less friction and much wear resistance in sliding components like bush bearings and linear sliding surfaces in combination with metal and polyamide.

Polyvinyl chloride (PVC): This thermoplastic can be modified for a number of properties. It is commonly used for coating electric wires, production of sheets and foils, pipes and pipe fittings and injection moulded components for agricultural and electrical gadgets. This plastic is flame retardant and difficult to process because of its sensitivity to heat. It chemically gets degraded and releases hydrochloric acid when exposed to heat for a longer period of time.

Polyethylene terephthalate: This is a clear and tough plastic most suitable for beverages and food products. Jars and bottles can be blow moulded by process stretch-blow moulding. Suitably dimensioned and shaped preforms are first injection moulded and then stretched-blow moulded.

Processing

Injection Moulding

In principle injection moulding of plastics is a process where plastic granules are melted by application of heat. Heat is applied to granules in such a way that it reaches to moulding temperature. Moulding temperature is that best temperature where plastics become soft enough or syrupy to be pushed forcefully and with speed into the cavities of mould. Once it is in mould cavities, starts cooling down and allowed to cool down to such an extent that plastics retain the shape of cavity. Components or mouldings are then taken out of cavities. Mouldings so taken out undergo further and slow cooling down to ambient temperature. During cooling down, dimensions of mouldings become stabilised and these should be required dimensions.

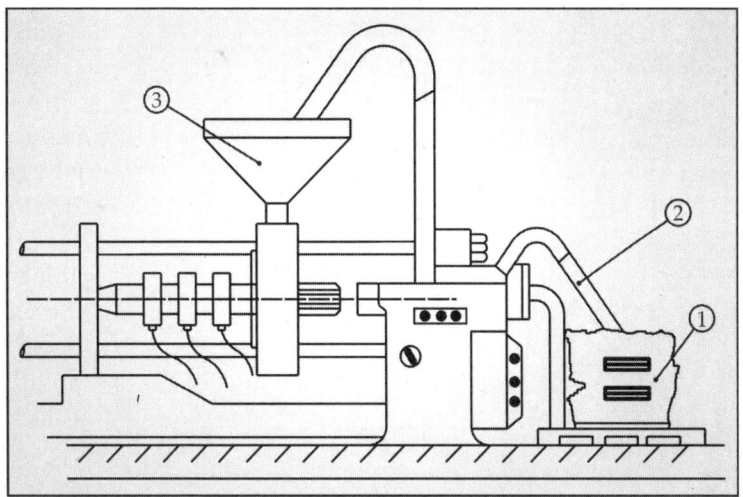

Fig. 1.1: Hopper Loader

Injection moulding of plastics is carried out with the help of a complete system. A typical system is shown in Fig. 1.1. ① is bag or a container containing plastic granules. It is placed on a wooden pallet on the ground. ② is sucking pipe of vacuum loader. Granules are sucked and fed to hopper drier ③ of injection moulding machine. There is suitable automation to control filling of hopper. As soon as hopper is filled with granules up to a predetermined level, loading system stops. When level of granules in hopper reaches a lower predetermined level, vacuum loader starts again. Granules from hopper reach directly over screw-plunger channel as shown in Fig. 1.2.

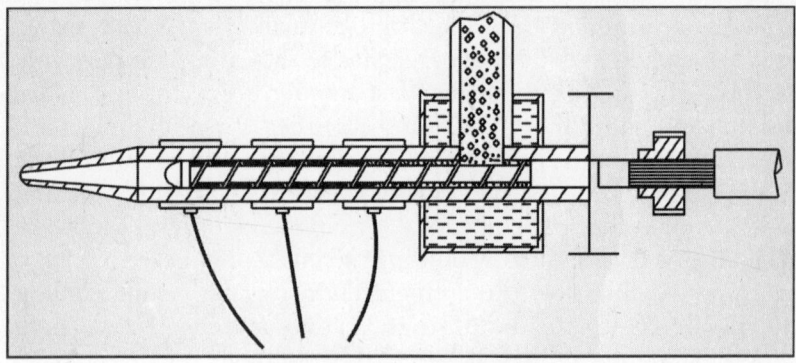

Fig. 1.2: Screw-Plunger

This portion of cylinder in which screw-plunger rotates and moves axially is kept cool so that granules do not get melted at this place to avoid formation of lump which prevents granules to reach over screw-plunger. As soon as screw-plunger rotates, granules in the spiral space of screw move towards nozzle, passing through various heating zones of cylinder. While granules are moving towards nozzle, get gradually heated and start offering resistance to movement forward towards nozzle. As a result of reaction, screw-plunger tends to axially move in the opposite direction, that is away from nozzle side. But this axial movement of screw-plunger is controlled by machine system, thus affecting the movement of plastic granules with a specified speed and pressure. While passing through various heating zones, granules take shape of homogeneous melt by the rotating action of screw-plunger. This melt keeps on getting collected in front of screw by passing through a non-return valve at the tip of screw-plunger. Melted and homogeneous plastic so collected between the nozzle and valve is now ready to be pushed out of nozzle. Figure 1.2 shows the stage when machine system is about to push melted plastics out of cylinder through nozzle.

In Fig. 1.3, machine system is in open stage (*a*) of moulding cycle. At this stage cylinder nozzle is away from mould feed head. Fixed platen of machine remains at its fixed position on which injection side of mould half is loaded. Moving platen of machine with other half of mould is away from injection side of mould half. When preset mould open time is completed and safety devices functioning properly, mould platen moves forward to close the two halves of mould with specified mould closing force. Now cylinder carriage moves towards fixed platen till nozzle's front surface touches mould feed head (*b*). Alignment of feed head and nozzle is kept such that holes of feed head and nozzle are co-axial. Contact of nozzle with feed head of mould is with a certain set pressure so that plastics melt does not leak out when forced by screw-plunger head to reach mould cavities through feed head, runners and cavity gates in the mould. For a predetermined and set time, machine and mould remain in 'mould closed' condition. This time allows for component in mould cavity to cool down to such an extent that no distortion or damage takes place to component during ejection of component from mould at the end of mould opening stroke. As soon as component ejection has taken place, next moulding cycle starts.

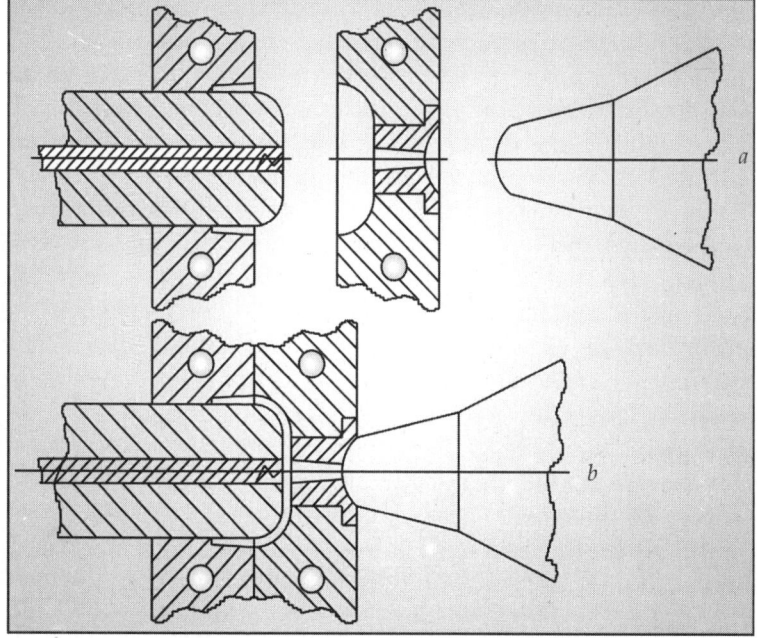

Fig. 1.3: Mould—Open and Closed

Quality Aspects

Quality aspects of injection moulded plastic components may be categorized as follows:

- Plastic material dependent
- Process dependent
- Environment dependent

Plastics material dependent, plastics material dependent properties are toughness, hardness, flexural hardness, surface finish, dimensional stability, clarity, light transmittance, colourability and mouldability.

Process dependent, process dependent properties are surface finish, dimensional stability, internal stresses and free from defects like weld line, flow marks, burn spots and silver streaks, etc.

Environment dependent, environment dependent properties are effect of light, effect of infrared and ultraviolet energy, effect of moisture and gases, effect of ambient temperature variation, effect of microorganisms, behaviour in vacuum, lack of gravity and high speed acceleration. Components or parts moulded from a tough plastic can stand impact and deformation without breakage or crack. Toughness varies according to type and grade of plastics. It is measured by a test known as Izod Impact test. Test values are given in ft-lb/inch of notch (1/8 inch thick specimen). Higher the value, tougher is the plastic. Tough plastics are usually used for moulding items like car bumpers, jerrycans, etc.

There are many items which need to give a feeling of hardness when held by hand like calculator frame, computer mouse and so on. There is a range of hardness. Components moulded from hard plastics like general purpose polystyrene may break if subjected to an impact, say, by accidental dropping. Advantage of a hard plastic is that it is scratch-resistant. Inexpensive items can be moulded. Clarity can be maintained. Transparent colour shades may be given. Items like glasses, dry fruit containers, small kitchen containers, backlight lenses and many other items can be moulded. Hardness of plastics is measured by preparing test pieces and subjecting them to testing procedure by a comparator machine or hardness tester. Hardness values may be given in Rockwell scale and Shore scale. Flexural strength is determined by a testing equipment. Test pieces are prepared to carry out testing of flexural strength. Test piece is subjected to flexing with a specified magnitude and frequency. After predetermined interval of flexing, test piece is inspected for development of traces of surface cracks. Number of flexing, with specified frequency, just before development of cracks, is flexural strength. Its unit is psi. There are automatic computerised testing equipments which give precise test results. Printouts may also be taken out or the results may be stored in computer memory.

Surface Finish

Surface finish of a moulded component, part or article depends on type and grade of plastics keeping the processing conditions, machine and mould the same. Parts moulded from high impact polystyrene will not have lustrous surface finish as compared to ABS plastics. Mat surface can also be achieved by preparing the surface of mould cavities by electric discharge machining or sand blasting. Figure 1.4 shows a typical component of an assembly where critical dimensions with tolerance are shown. These are designed dimensions. This component may be moulded in high density polyethylene (HDPE), polypropylene (PP) and acrylonitryl butadine styrene (ABS).

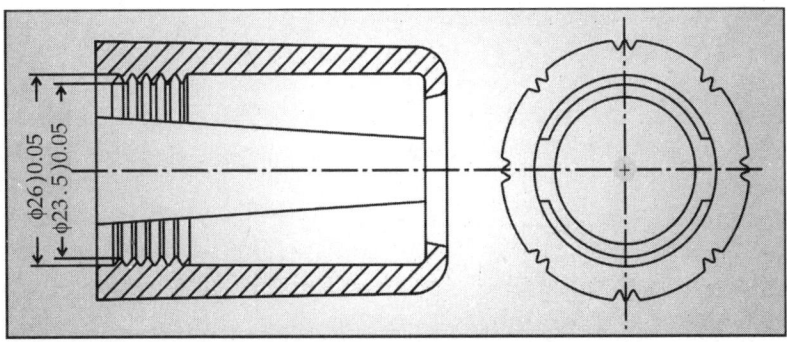

Fig. 1.4: Threaded Component

Above-mentioned plastic materials have different mould shrinkage values. These values are typically given in Table 1.1.

It can be noticed that mould shrinkage and its range is highest in case of HDPE. PP has medium mould shrinkage and its range. ABS has minimum mould shrinkage and its range.

S. No.	Plastics	Mould shrinkage		Range
		mm per mm	*%age*	*Per cent*
1.	HDPE	0.015–0.040	1.5– 4	2.5
2.	PP	0.010–0.025	1–2.5	1.5
3.	ABS	0.004–0.009	0.4–0.9	0.5

Table 1.1

Core and cavities of mould are designed and dimensioned in such a way that mould shrinkage allowance is provided so that specified dimensions are achieved once component is given time to cool down to ambient temperature after ejection. Still there would be variation in dimensions due to inherent range of shrinkage characteristics of each plastic. Table 1.2 shows specified and expected values of two dimensions shown in component drawing. From Table 1.2 it is evident that component produced from HDPE may have dimensions much out of specified tolerance. Component moulded in ABS has dimensions well near to limits of dimensions. It can, therefore, be appreciated that many a times consideration of shrinkage characteristics of plastics may be a deciding factor in selection of a particular variety and grade of plastic material. For details of testing methods, books on testing of plastics may be read.

Table 1.2

S. No.	Plastics	Mould shrinkage range	Specified dimension		Expected dimension	
			Min	*Max*	*Min*	*Max*
1.	HDPE	0.015–0.040	25.95	26.05	25.7	26.3
2.	PP	0.010–0.025	″	″	25.8	26.2
3.	ABS	0.004–0.009	″	″	25.93	26.06

Injection moulding process parameters are of paramount importance for the dimensional stability in moulded component, especially in bulk production. Injection moulding process parameters of importance are as follows.

- Homogeneity of melt in heating cylinder of machine, especially the melt in front of screw-plunger, ready for injection

- Melt temperature and its range of variation in bulk production
- Dryness of plastics melt. That means extent of presence of moisture and vapours
- Speed of injection of plastics melt in cubic centimetres per second
- Design and polish of feed head, runners, gates and cavities
- Mould cavities temperature with extent of variation
- Heat transfer rate from cavities and core surfaces to circulating cooling medium
- Injection pressure and speed profile. This means pre-determined variable pressure and speeds at various stages of injection
- Back pressure
- Cooling time
- Effectiveness of mould temperature control
- How stabilized are complete moulding cycles in terms of pressure, speed, temperature, time, etc.
- Availability of computerised self-diagnostics and corrective actions on cycle to cycle basis
- Automatic corrections of process parameters, necessary due to variation in environmental conditions. This means ambient temperature, humidity, extent of dust, fumes or any other gases
- Maintenance of general cleanliness and handling of moulded components just after ejection from machine till dimensionally stabilized

Precise control over temperature, pressure, speed and time requires sensitive and quick response sensors. Design of machine elements, hydraulic circuitry, etc. should match with quick response sensors. Microprocessor based controls take care to compensate for inherent delayed responses. Further, preset process parameters for various components can be quickly set by pressing a few keys on microprocessor control panel. Sturdiness of machine is important so that no undue deflection in platens takes place, thus avoiding formation of flashes on parting faces of mould. All machine system, especially safety related should be highly reliable.

Mould temperature control units are usually separate from machine. But in sophisticated machines, mould temperature control and actual pressure in cavities are also controlled through microprocessor.

Components for optical purposes, like lenses, reflectors of cars, health care equipments are moulded in a shop where clean air with definite humidity range and ambient temperature is maintained. Atmosphere in a shopfloor should be as free from viruses, bacteria and germs as possible. Automation for packing of components is desirable so that human touch should be as less as possible.

Injection Moulds

Design and construction of injection moulds in itself is a vast subject. Coverage of this subject in great details is beyond the scope of this book. However, some basic knowledge is provided in coming paragraphs. Following are the types of moulds:

- Two-plate
- Three-plate
- Single and multi-cavities
- Injection of plastics melt to cavities through feed head, runners and a variety of gates
- Pin point gating with hot runner system, scrap is almost eliminated
- Self-degating system
- Mould with automatic stripping or unscrewing system
- Multi-layer or colour moulding system in mould. Indexing type
- Having hydraulically withdrawing cores
- Design suitable for automatic switching over of moulds in machine from mould conveyor

Figure 1.5 shows 'two-plate' and 'three-plate' moulds. In two-plate mould, plates ① and ② are fixed whereas in 3-plate mould, plate ② is floating over guide pins ③. Choice of selection of 2 or 3-plate mould very much depends on component design, dimensions and finish required. In the same figure single and multi-cavity moulds are also shown. Single cavity mould may be that for a bucket having a direct feeding of plastics melt through feed head. Moulded item would come out with sprue (a small tapered length of plastics formed in feed head of mould). It is removed by an auxiliary operation.

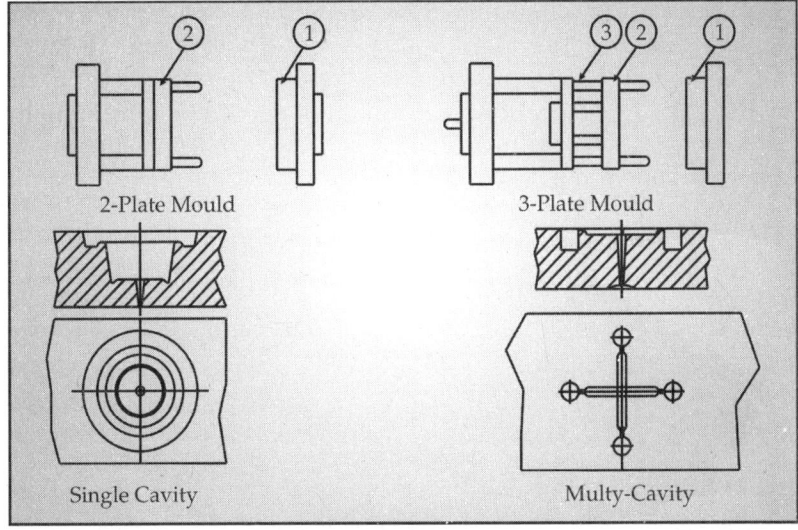

Fig. 1.5: Types of Moulds

Various types of cavity gates are shown in Fig. 1.6.

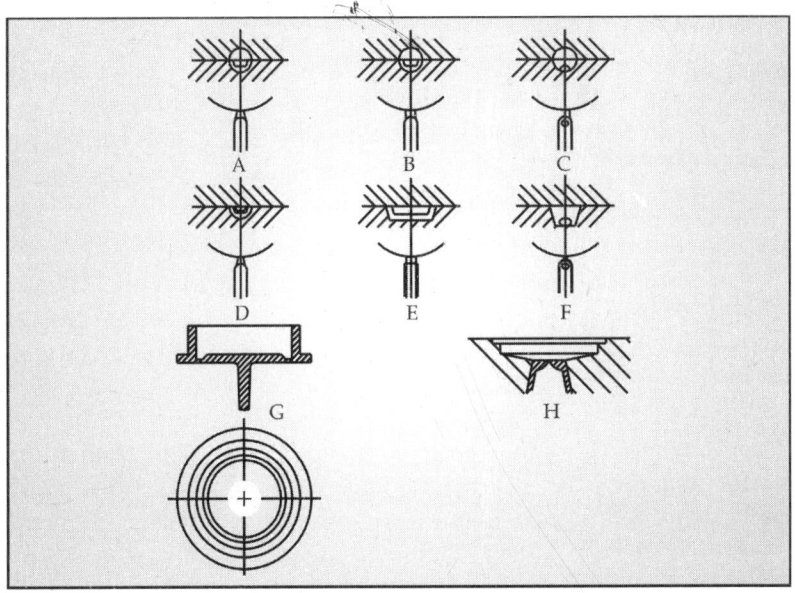

Fig. 1.6: Cavity Gate

Multi-cavity mould is for a component, say a simple lid of a bottle where injection point at the brim is permissible. In case injection point (gate) would not have been permissible at the brim then the design of mould would be different. Plate ① would then be floating and runner would have been on the back. Such type of runner is shown in Fig. 1.7.

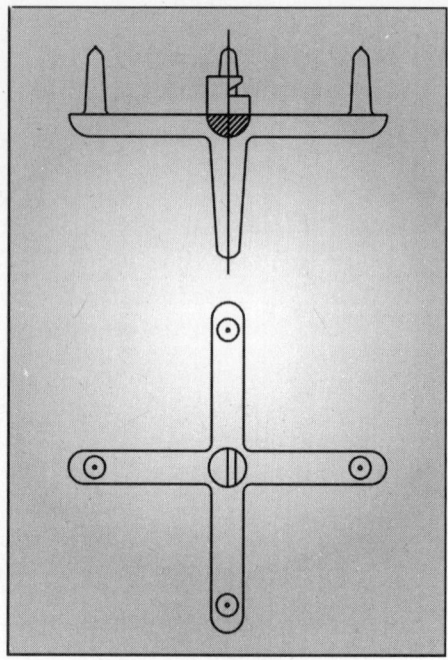

Fig. 1.7: Runner

In this type of mould removal of scrap is difficult. In case generation of scrap is not desirable, hot runner, pin point gating system would have to be employed. Such moulds are quite expensive and need careful use and handling.

Design and quality of two complicated components are described here. Working principle of moulds is also explained for better appreciation of intricacies involved in design and construction of mould.

Figure 1.8 shows three components of an assembly. ① is already existing metal body with right hand 1.25 mm pitch threads. ② is the 'head' of assembly which is to be screwed over the body. On the larger diameter side of head, a ring ③ having 2.5 mm pitch threads (right hand) would be screwed.

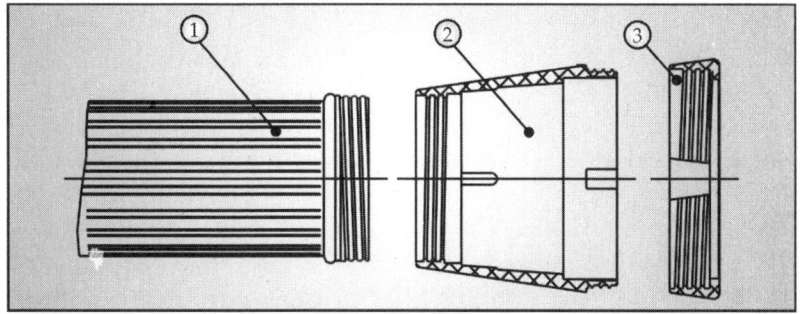

Fig. 1.8: Matching Components

A mould was to be designed and made with following functional requirements.

- Mould should have two cavities
- Component should be free from possibility of weakness due to weld line
- Both the components should automatically be unscrewed and removed from mould
- There should not be any parting line on external or internal threads
- Component would have to be moulded in High Impact Polystyrene (HIP)
- Mould should not be expansive
- Mould should be easy to maintain

Scrap generation should be as less as possible.

With above design constraints, designer (author of this book) started thinking and visualizing various design options. Making of hot-runner mould was ruled out due to high cost and lack of experience with such moulds among technicians and operators. Use of collapsible core for inner threads and opening jaws (plates) for outer threads was also ruled out as no flash was tolerable. The only option left was removal of component by unscrewing from threaded core as well as from threaded cavity. Adoption of this technique was not possible unless both threads, inner and outer have same value of right hand pitch. Matter was discussed with technicians and administrator of sheet metal shop. Change of thread pitch of sheet metal body was ruled out. So it was decided to change the pitch of thread in ring to 1.25 from 2.50 mm.

Experimental mould was made to mould component by keeping gate at the brim of larger diameter. This had not given satisfactory strength at the weld line. Hence some components got cracked at weld line in drop test. It was then decided that a disc gating would be the best option, in spite of the fact that percentage of scrap in the form of disc was high. But in the interest of reliability, strength of component, high percentage of scrap was tolerated. Weight of one component, head ② was about 28 grams and that of cut off disc gate, 1.8 grams. Hence percentage of scrap was 6.4. Basic design of mould was as shown in Fig. 1.9.

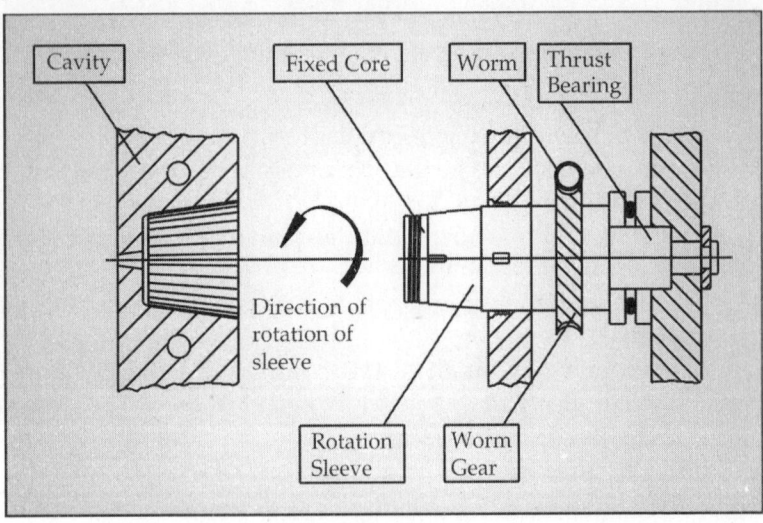

Fig. 1.9: Disc Gating and Auto Unscrewing System

Component came out of mould was with disc gate as shown in Fig. 1.10.

It was then loaded on a rotating jig on a lathe and disc was cut off to have a clear start of thread.

Another example is of a complicated component weighing about 135 grams in polypropylene copolymer. Basic design of component is shown in Fig. 1.11.

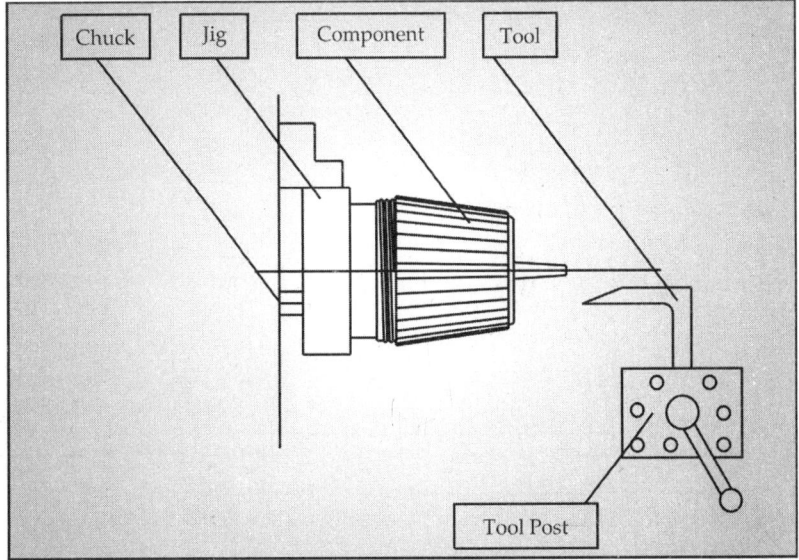

Fig. 1.10: Degating Jig

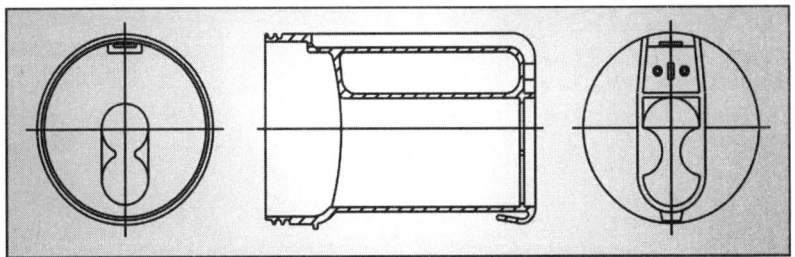

Fig. 1.11: Basic Component Design

A single cavity mould was designed to be loaded on a fully automatic, 250 tons mould locking force. Selected machine was having two optional hydraulic ports to operate two cores independently. Sequence of complete moulding cycle, including core withdrawal and repositioning could be programmed on microprocessor control console. Basic design of mould is shown in Fig. 1.12.

In Fig. 1.12 parts of mould shown are sturdy mould plate ①, core ③, auxiliary hydraulically operated core ④, cam pin operated top core ⑤, hydraulic jack ⑦, injection sprue cum gate ⑧ and cooling circuit nozzles ⑨.

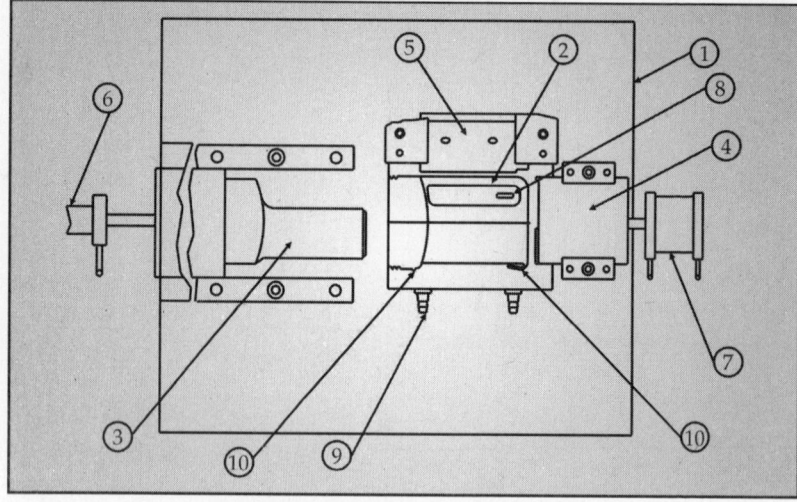

Fig. 1.12: Basic Mould Construction

Sequence of machine and mould functions is as below:

- Mould is in open condition. Core ③, ④ and ⑤ are in withdrawn position
- Start command push button on microprocessor control panel is pressed by operator
- Moving platen of machine starts to close. Simultaneously main and auxiliary cores move in to touch each other at the travel limits
- Mould halves approach to close. Some distance before actual closing of mould, cam pin operates top core ⑤
- On final closing of mould all the three cores are locked in position
- Mould is now closed with pre-adjusted mould locking force
- Injection of melted plastics takes place and back pressure holding time is given for plastics in gate to solidify
- Mould remain closed for pre-set component cooling time
- Mould starts opening and stops midway. Hydraulically operated cores withdrawal is completed
- After withdrawal is completed, mould moves to its extreme open position
- Ejector pins operate to push the component out of cavity
- Moulding cycle repeats by itself if machine is set to fully automatic operation

Blow Moulding

A large number of varieties of articles like cold drink, mineral water and beverages, bottles, small and big jerrycans, cosmetics, medical, agriculture and technical items are produced by a blow moulding process.

Blow moulding is carried out by different methodologies, which are as follows:

- Continuous extrusion blow moulding
- Intermittent extrusion blow moulding
- Injection-blow moulding
- Pre-formed stretch blow moulding

Figure 1.13 shows a typical continuous extrusion blow moulding system. Extruder ① continuously feeds melted plastics to extrusion head ②. This head delivers melted plastics in vertical downwards direction in the form of a tube ③, which is called parison. Diameter, wall thickness of parison depend upon shape and adjustment of extrusion head, vertical continuous delivery of parison with a pre-determined and set linear speed in centimetre per second.

Underneath the extrusion head there is a blow mould handling unit ⑤ which also carries parison cutting knife and high pressure air blowing nozzle. Handling unit moves up and down, mould ④

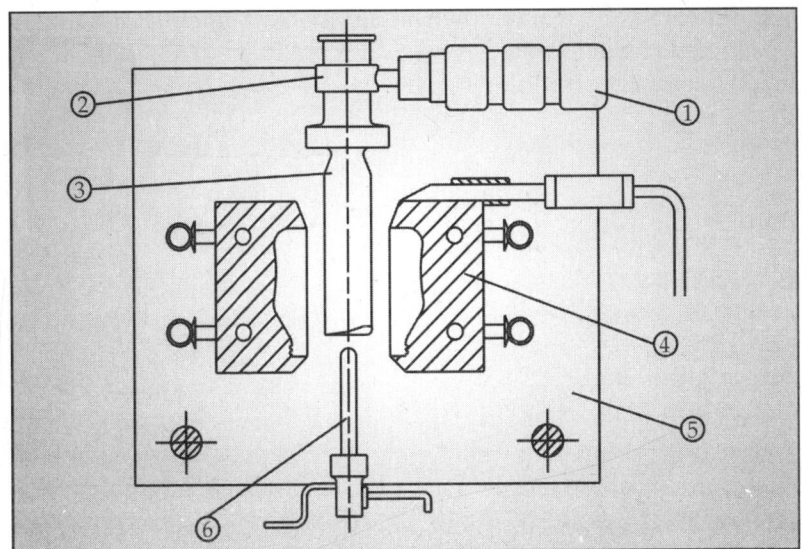

Fig. 1.13: Blow Moulding Setup

closes and opens, high pressure air nozzle/pipe ⑥ also moves up and down in relation with mould. Cutting knife also operates to cut parison. During continuous blow moulding process the following parameters are accurately controlled by sensitive sensors and close loop circuits.

- Parison diameter
- Parison wall thickness
- Parison temperature
- Parison downward speed

Sequence of operation is as under.

- Mould carrier is in up position
- Mould is opened enough to allow parison to move down between the two halves of mould
- Air nozzle is in up position in relation with mould
- Parison moving down with a specified and set speed
- After a specified length of parison comes out of extrusion head, mould system starts moving down with parison speed
- Simultaneously mould closes
- Simultaneously parison is cut
- Mould gets shifted to one side, enough to make room for continuous parison extrusion
- Mould opens and blown article is ejected together with scrap (flash) inherent to process
- Immediately after ejection, open mould moves back to take next length of parison
- Process then repeats itself

Materials like LDPE, LLDPE, HDPE, PP, HM HDPE can be extrusion blow moulded. Extrusion blow moulding process is also suitable for PVC. PVC is very sensitive to heat. It can degrade rapidly if gets overheated.

Intermittent Extrusion Blow Moulding

This technique is generally used for large containers like twenty five litre jerrycan. Figure 1.14 shows a typical arrangement of intermittent (non-continuous) extrusion blow moulding.

- In closed condition, axis of mould cavity ⑦ is coaxial with axis of parison extrusion die ⑤

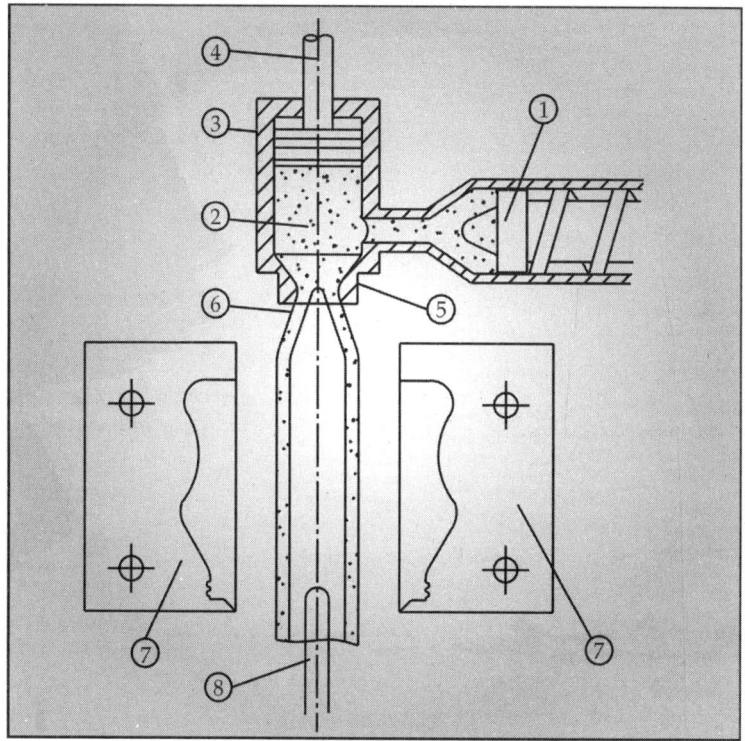

Fig. 1.14: Non-Continuous Parison

- There is no vertical movement in mould. It only opens and closes
- There is a hollow vertical pin ⑧. Its axis coincides with axis of mould cavity 'mouth' opening (threaded opening) in closed condition. Its function is to inject high pressure air or nitrogen (as the case may be)
- Mould closes completely with parison inside
- Extruder ① is continuously feeding melted plastics in accumulator chamber ②
- Plunger ③ (operated by rod 4) inside chamber pushes melt down with a specified and set speed through extrusion die to form a parison
- Parison on the top is cut by knife
- Moulded item cools down to ejection temperature
- Mould opens and blow moulded item is ejected

- Next parison immediately arrives between the open halves of mould
- Process then repeats itself

Items come out with flashes which can be removed manually or by automation. In case of automation, blow moulded article with flashes is automatically picked up from ejection point and transferred to another station where de-flashing takes place. In case of extrusion blow moulding process it is possible to mould articles with handle formation. Limitation with this process is that neck (opening) dimensions and shape cannot be maintained with precise tolerances. Sophisticated blow moulding items for technical purposes are generally required to have uniform wall thickness with close tolerances. Sometimes shape of article is such that blowing of parison varies greatly. Wall thickness remains thin where blowing is more and thick where blowing is less. To overcome this deficiency, to a great extent, sophisticated machines are used where variable thickness of parison wall can be extruded. It is controlled by microprocessor. Predetermined 'profile' of wall thickness is set on computerized control panel which activates extruder's die mechanism to generate different wall thicknesses at different portions of full length parison.

Injection Blow Moulding

As the name suggests first melted plastics is injected into a cavity to form a moulding with precision neck having threads or locking design. Remaining body of injection moulded component still kept hot on the core is transferred to another blow moulded station.

As soon as mould closes, high pressure air from core passage enters to blow the injection moulded part to take shape of blow moulding cavity. Blow mould opens and core system indexes to reach third station where ejection of injection blow moulded article takes place. Figure 1.15 shows basic system of injection blow moulding machine.

Indexing turret carries three similar cores of suitable design and dimensions. There is arrangement to sense and maintain predetermined temperature. Also, all the three cores have air valve system to inject high pressure air. There are following stations:

- A. Injection of plastics melt where moulded thread or locking portion is quickly cooled down. But portion to be blown on next station is kept at blowing temperature.
- B. Mould opens and turret gets indexed.

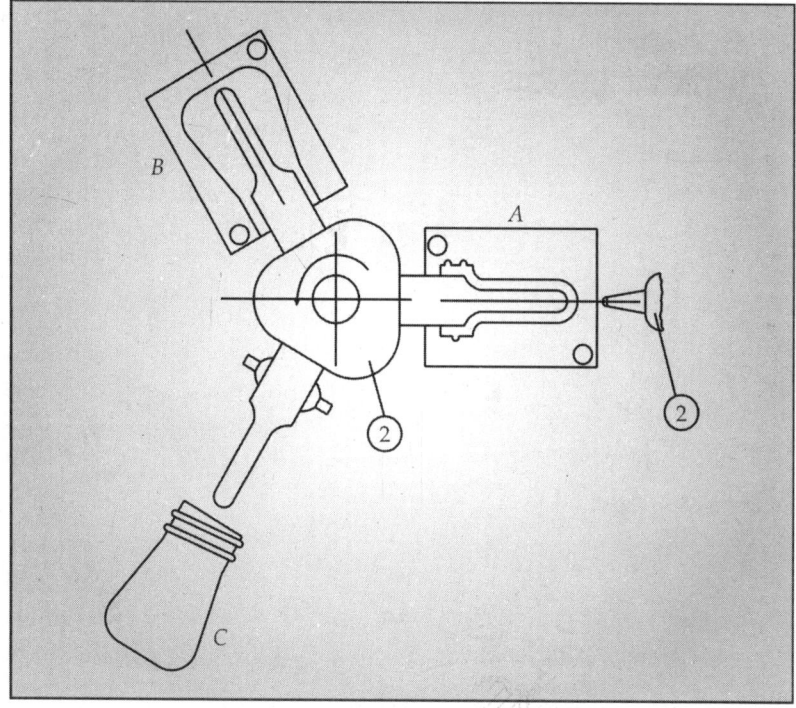

Fig. 1.15: Injection Blow Moulding

At this station a blow mould closes over the already injection moulded preform. High pressure air or nitrogen is injected through core to blow the preform to take shape of cavity. Blown item is cooled down in cool cavity. Mould opens to provide enough space for blown item to move out of blow mould when turret gets indexed to station C where blow moulded item is ejected. So each indexing gives a component. Injection blow moulding can also be done in another way. Preform are first moulded and collected. They are then heated to stretch blowing temperature and fed to stretch blowing unit. In this unit preform is stretched by means of a rod and then blown. This process is specially used for producing PET bottles. Stretch blow moulding improves mechanical strength of bottles, improves surface finish and gas permeability. With injection blow moulding process, handled containers cannot be produced.

As said earlier, precision threads or locking system in product is produced with precision.

Thermoforming

This process is used to produce shallow or deep articles like glasses, food service trays, inner lining of refrigerators, containers, large panels, dash boards of automobiles and a variety of packing items.

Following forming techniques may be considered under thermoforming process:

- Vacuum forming
- Draw assisted vacuum forming
- Vacuum-cum-pressure forming

Basic principles of above techniques are briefly described with the help of typical Fig. 1.16.

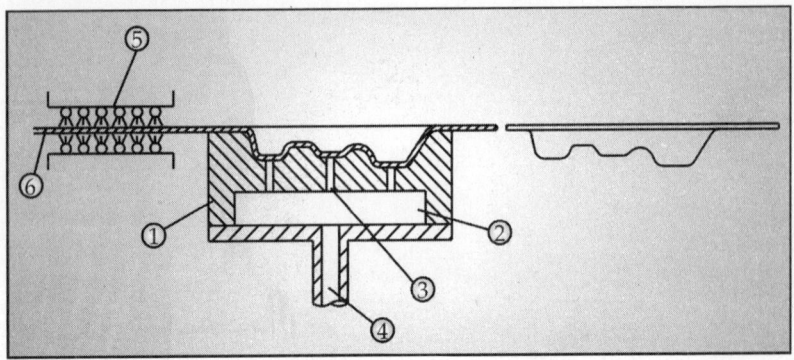

Fig. 1.16: Thermoforming Setup

① is cavity block having a machined and polished cavity. Shape and dimensions are so kept as to give thermoformed article of designed shape and dimensions. Cavity has a number of small through holes which open into a vacuum sealed space ②. A pipe ④ is connected to sealed space to create vacuum in cavity through a number of holes, connecting sealed space to cavity. Plastic sheet which is to be vacuum formed is heated on one or both sides by means of suitable heaters ⑤. Heating of sheet ⑥ is done to such a temperature and time that it becomes soft to formable condition. At this stage, sheet is automatically transferred over the cavity and vacuum is created fast inside the cavity. Consequently, sheet is sucked to take shape of cavities. Once formed, formed sheet is transferred to trimming station where brim is trimmed according to trimming tool which is designed and made to give desired trim profile and dimensions.

Heating of sheet can also be carried out by hot air. This depends on type, grade and thickness of plastic sheet. High impact polystyrene, acrylic, PVC, linear polyethylene, polypropylene and cellulose acetate are some suitable plastic sheets for thermo-forming.

Draw assisted vacuum forming is used where deep forming is required to be done. Basic principle of deep vacuum forming is briefly described with the help of Fig. 1.17.

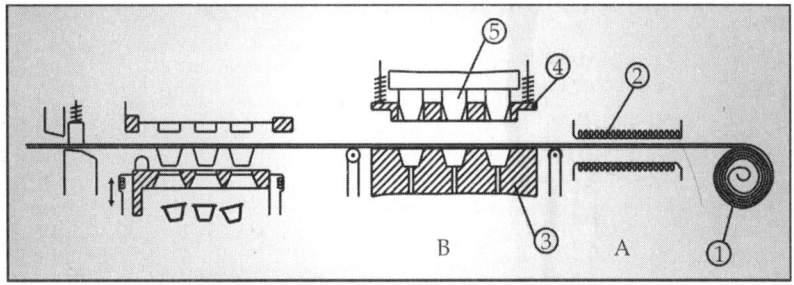

Fig. 1.17: Thermoformed Container Line

Thin-walled cold drink glasses are produced by multi impression high speed system to produce bulk quantities in a given time, say fifty thousand pieces in a shift of eight hours. ① is roll of plastics strip of thin gauge, say 0.9 millimetre. It passes through station A, B, C and D. On station A, intermittently travelling strip is heated by heaters ②. As soon as heated portion of strip reaches over mould ③ at station B, holding pad ④ and draw punches ⑤ come down. Pad keeping the strip pressed and draw punches to push the heated strip inside the cavities of mould to a certain depth and at that moment vacuum is applied to cavities by which heated sheet takes complete contours of cavity. At the same instant chilled cavities cool the formed glass (container) shapes. Sheet with formed glasses is lifted up to clear it of the mould surface for further transfer to station C where trimming takes place and trimmed glasses fell down on collecting chutes or conveyor. For some thermoforming operations, where the thickness of sheet is more, to make it stiff, high pressure air is also used to support vacuum formation. This results in accurate formation of contours of cavity. High pressure air is injected through pusher punches as soon as vacuum in the cavity is created. Basic arrangement is shown in Fig. 1.18.

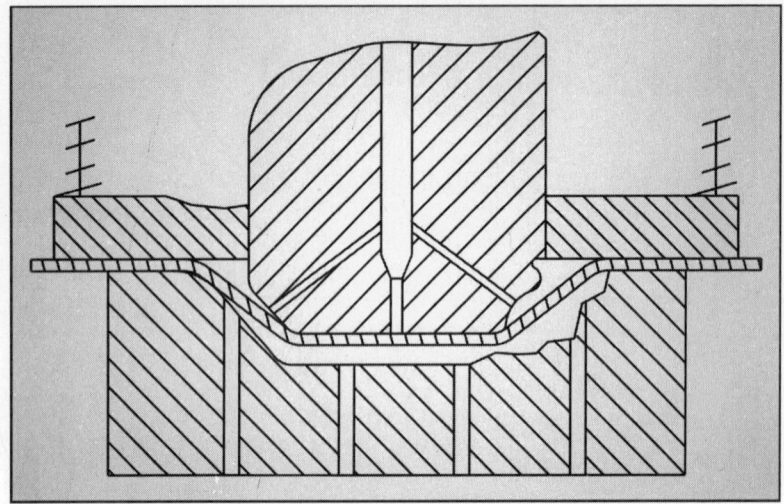

Fig. 1.18: Pressure-aided Thermoforming

Photographs

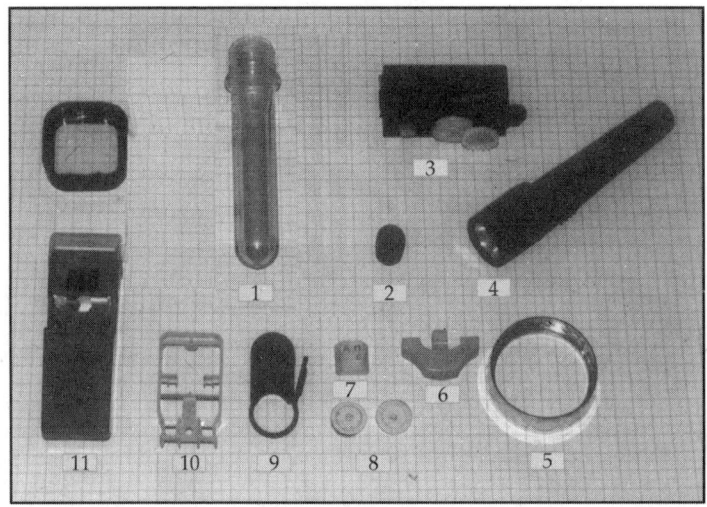

Photo 1.1

Components

1. It is a preform of PET for blow moulding of bottle.
2. It is a component which is hollow and has threads on diameter for one-third of length.
3. A drive mechanism of a toy car. It has a number of gears and other components. A gear and pulley of white colour plastic can be seen.
4. It is a complete flashlight operated by two penlight cells (2 × R6). It has precisely designed and moulded components of HIP or ABS plastic. Reflector is moulded in general purpose polystyrene and vacuum metalised.
5. It is a ring having internal threads. Threads are moulded on a collapsible core for easy ejection. Component is moulded from plating grade ABS and is electroplated.
6. Sliding switch having pair of legs of snap assembly design.
7. Key of a typewriter. It has printed alphabets in two colours.
8. These are two white colour gears having very fine teeth.
9. A snap locking type of key or tag ring. It is very precisely designed and moulded component.
10. Another component having complicated and precisely dimensioned sizes.
11. It is a complete handy flashlight having a number of components like body, cover and bulb holder which is moulded in ABS. Front is a clear half round type of cover. It is moulded in acrylic.

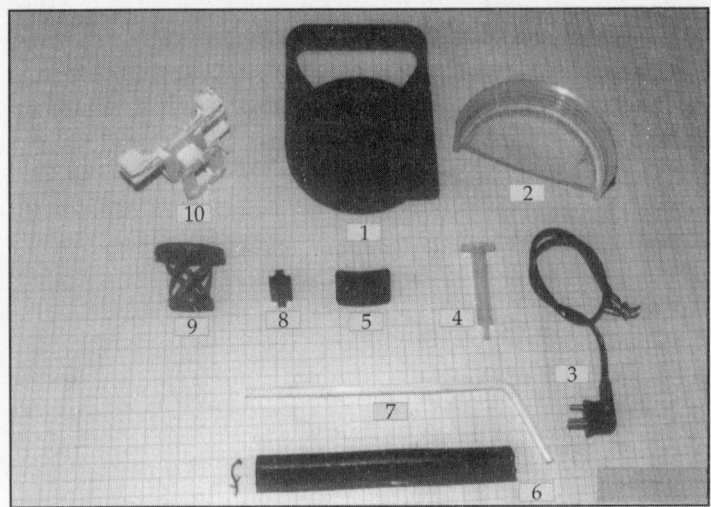

Photo 1.2

Components

1. It is cover-cum-component holder of an assembly, has a number of ribs, legs, screw holes and undercuts, weighs about 38 grams and moulded in ABS.
2. Is a cut section of a lid (generally called headring) of a flashlight lantern. It has serrations on diameter and threads inside. Component is ejected after automatic unscrewing while component is still in floating cavity and mould is partly opened.
3. Insert moulded precision three pin plug.
4. Is a cylinder of 5 ml throw away syringe, moulded in polycarbonate. Dimensional quality of bore is very precisely moulded.
5. Sliding switch rider, moulded in HIP (high impact polystyrene).
6. It is a cut piece of a long extruded section of PVC.
7. Tiltable straw of polypropylene or HDPE. Bending portion has specially designed corrugated circular rings. These rings are generated by specialized automatic machine.
8. A piece, cut from a circular extruded length. It has formation on diameter and there is hole through out the length.
9. Is a cap cover for a tablet bottle/phial. It has a ring attached to cover with three helix-shaped legs. The whole thing is moulded in one piece. It has a springy action in legs. Purpose is to keep cotton slightly pressed over the tablets.
10. It is a 3-dimensional complicated component moulded in white colour ABS. A number of ejector pin marks and parting lines are there.

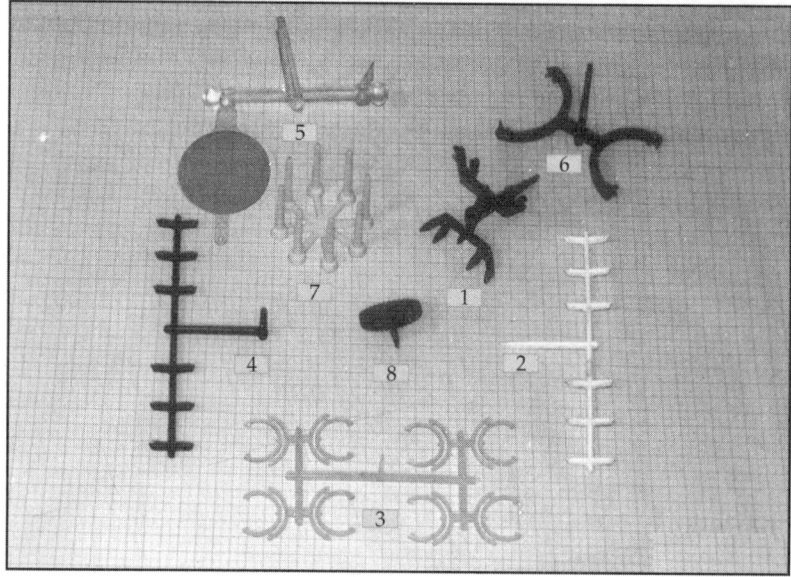

Photo 1.3

Components

1, 2, 4, 6 and 7 are the runners having pinpoint gating. At the time of mould opening, component snaps out of die and runner is pulled out automatically, thus br'nging out formed inclined runner near the pin point gating.

3. This is a runner of very unique design. It is good example of well-balanced runner and gating design. There are three gates at 120 degrees apart, opening in the cavity. These three gates are fed from equidistant two points and these two points are fed by runner which is equidistant from two points. In this way flow of plastic is well balanced. This results in providing condition to get good quality component.

5. A PMMA lens attached with its runner, 'Fan' gate and sprue is shown.

8. It is a gear having a taper protrusion (axil). Gate is just on the small round face, directly from sprue.

2

Press Working of Sheet Metal

Properties of Sheet Metal

A variety of items like automotive parts, components required in electrical, electronic gadgets, components used in telecommunication system, kitchen wares and utensils, etc. are produced out of metals. Metals, in the form of sheets, strips and blanks, generally used are brass, aluminium, stainless steel, iron, and tin coated iron sheets. Zinc, phosphor bronze and other specialized materials in the form of sheets are also used. Metallurgical advancements have given rise to metals of specialized composition of different metals and elements. These are converted into the form of sheets, strips and blanks. These are used to produce components which are to perform critical function in an assembly of system. Components may be tiny but critical throughout their service span of time. Example of such components can be found in systems of spacecrafts, aircrafts, weather advance warning system, safety and security alarms.

Properties of sheets also depend on the manner in which metals are rolled to form sheets of desired thickness with specified tolerances, width, hardness and grain direction due to rolling process.

Following properties are normally taken into consideration while designing components.

Tensile strength, compressive strength, hardness, ductility, density, shear strength, grain direction, and surface finish, etc.

Press working of sheet metal means that sheets are given a variety of shapes and sizes with the help of suitably designed tools and correctly selected presses. Presses may be hand or power operated. Hand presses may be manually operated, pneumatically or hydraulically operated. Small capacity presses are used for producing tiny components. These are also used for assembly work like riveting. Heavy mechanical and hydraulic presses are used to perform various operations of press working of sheet metals. Some commonly used operations are listed below.

- Cutting and shearing
- Blanking
- Trimming
- Bending
- Forming
- Stamping
- Drawing
- Deep drawing
- Piercing
- Beading
- Notching
- Thread rolling

Photograph 2.1 shows a number of sheet metal components to visualize as to what operations and its sequences are required to produce shown components. Sequence of operation for a particular component can be well visualized once the concepts of various sheet metal working operations are understood. Therefore, various operations are briefly described in following paras.

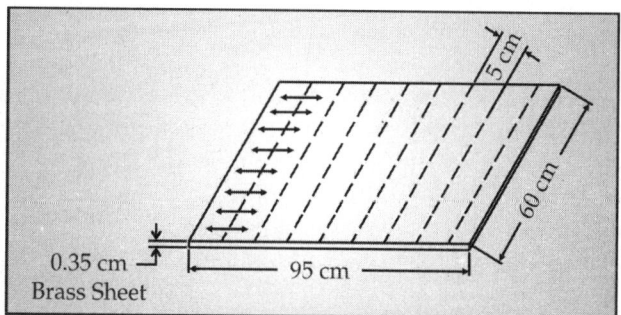

Fig. 2.1: Grain Direction

Cutting and Shearing

On most of the occasions, it is necessary to cut small strips from stock of large strips or rolls.

In Fig. 2.1, a typical brass sheet of 60 cm width, 95 cm long and 0.35 mm thick is shown. Grain direction, due to rolling, is shown by small arrows parallel to edge of sheet. Strips of 5 cm width are required to be cut. It is worth noting that cut strips would have grain direction at right angle to long edge of strips. For cutting one or two pieces, pair of scissors can be used. But it is not a feasible proposition for production of large number of strips. For this purpose, shearing machines (guillotines) are used. Figure 2.2 shows a guillotine in action.

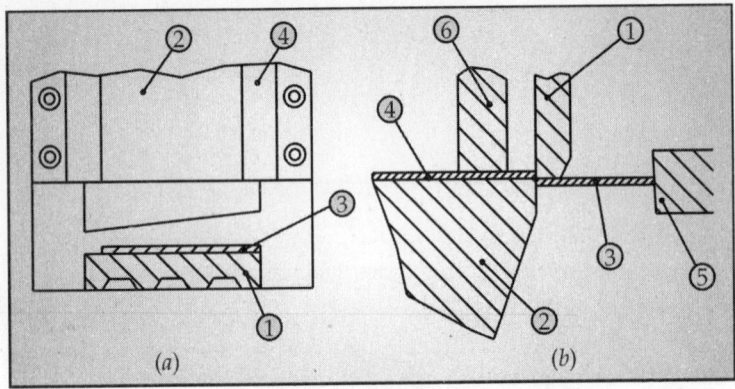

Fig. 2.2: Guillotine

Figure 2.2 (*a*) shows basic concept of a guillotine. ① is a big cast iron bed of the machine. Big sheet is placed over it. ② is upper moving shear blade which has an inclined cutting edge having an inclination of few centimetres. ③ is lower shear blade which is suitably fixed to machine bed ①. While upper shear is in up position, there is a gap between two shears. Sheets to be cut is pushed forward through the gap to touch a stopper. Stopper is so adjusted that required width of cut strip is obtained. Figure 2.2 (*b*) shows shearing action in progress. Pressure pad ⑥ presses the sheet before shearing action starts. Shearing blade ① travels down and shears the sheet to strip ③ while lower fixed shear supports the sheet.

Inclination in cutting edge of upper moving blade is provided so that shearing takes place gradually. This is to avoid sudden shock load on shearing blades and machine system. Guillotines are designed and built for a maximum thickness of sheet metal and maximum and minimum width. Force required to cut strips from a sheet may be calculated by the following formula.

Force = Cutting area × Shear strength of material

$F = A \times fs$ where F = Force required in tons

A = Square centimetres

fs = kg/cm^2

Area A is the product of sheet thickness and effective shearing length while inclined shearing blade is shearing the sheet progressively.

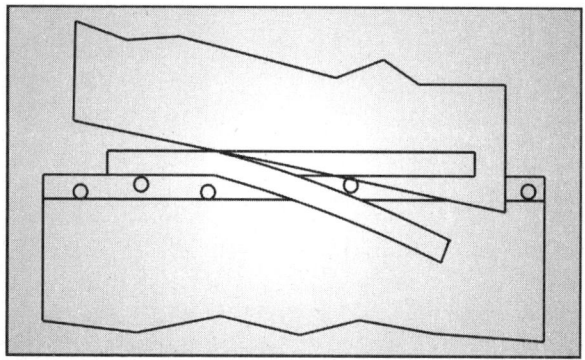

Fig. 2.3: Shearing Action

If there is no inclination in upper shearing blade, cutting area would be,

A = Sheet thickness, t × width of sheet, w

This would cause sudden and jerky load on the machine. Many a times it is necessary that grain direction should be along the length of strip. In that case sheets are placed on the guillotine table in such a way that grain direction are parallel to shearing blades. Consequently, cut strips would have grain direction along the length of strips.

Blanking

Figure 2.4 shows a phosphor bronze component and its blank which is 6.80 mm wide having a tolerance of ±0.10 mm. Strips of 6.8 mm width can be cut on guillotine but a close tolerance of ±0.10 cannot be ensured. Therefore, it is necessary to produce this strip by a blanking tool. Basic design and construction of a typical tool is shown in Fig. 2.5.

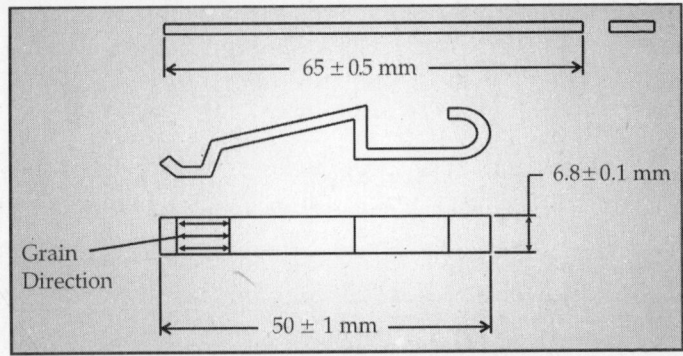

Fig. 2.4: Component and its Blank

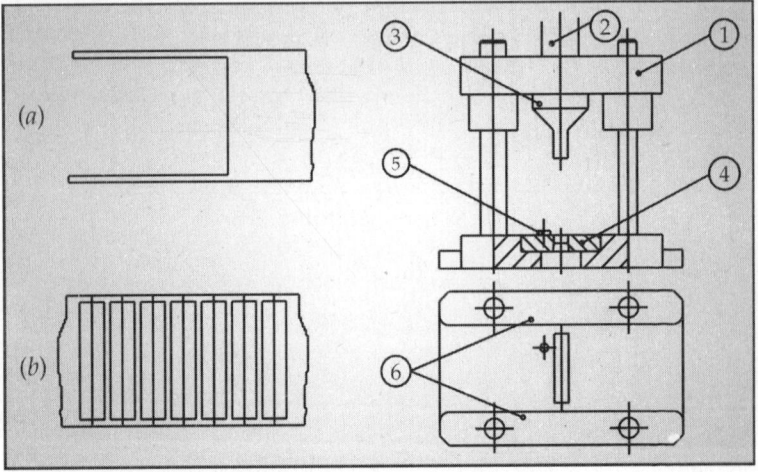

Fig. 2.5: Blanking Tool and Leftover

① is pillar type dieset. It is fitted with stem ②, blanking punch ③, blanking die ④, stopper ⑤, and guide plate ⑥. Distance between guide plates is so maintained that strip of specified dimension and tolerance can move freely with minimum possible shift along width. Blanking punch ③ and die ④ are well matched with a cutting clearance of about 0.05% of sheet thickness for phosphor bronze. Cutting clearance finely varies with sheet material and thickness. Cutting clearance for steel of 0.36 mm thickness would be 0.06 to 0.08 mm, depending upon shear strength of strip material.

Blanking principle is explained in the following paragraph with the help of Fig. 2.6. It shows a blanking punch passing through strip and then in blanking die. A perfect straight cut edge is shown. This does not happen. What actually happens can be seen in Fig. 2.6 (*b*). Blank is sheared out of strip. Shearing action is in a slant manner as shown. Force required to punch out a blank is calculated for a typical zinc blank.

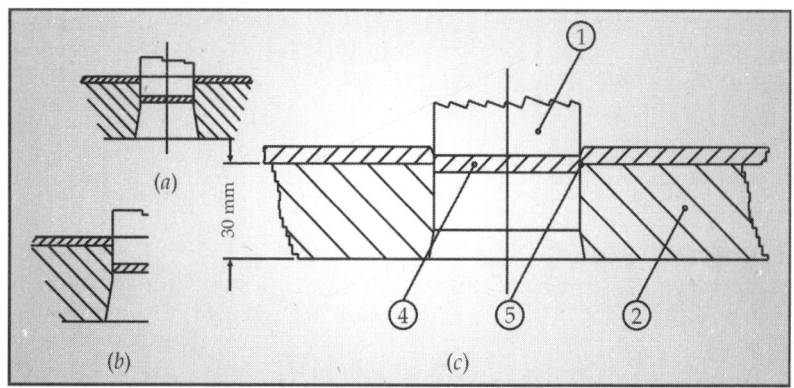

Fig. 2.6: Shearing Action in Blanking

Formula for calculating blanking force is

$$F = \text{Peripheral length} \times \text{thickness} \times fs$$
$$= 1.2 \times 6 \times 0.32 \times fs$$
$$= 2.34 \times fs \text{ where } fs \text{ is shear strength in kg-}f/\text{cm}^2$$
$$= 8225 \text{ kg}$$

Shear strength, fs for zinc is taken as 3515 kg-f/cm^2, therefore, theoretical value of force F required to blank out one piece works out to be 8225 kg.

In practice blanking tools with five or seven impressions are generally used. So five or seven blanks are obtained in one stroke of power press. For a five impression blanking tool, theoretical force required would be 5 × F which equals to 41 tons. In practice, a press is selected which has a tonnage at least five times the theoretical force required. Hence a power press of 200 tons is required to blank five blanks of zinc, as shown in Fig. 2.7. Power press of standard tonnage are selected. Use of power press of still higher rating is preferred because wear and tear in power press would be much less as total blanking load is much less as compared to power press tonnage.

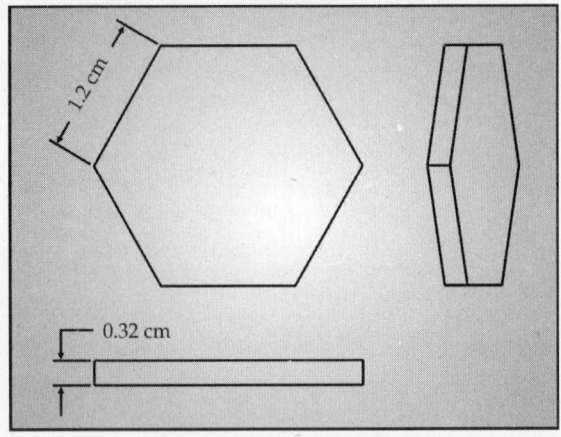

Fig. 2.7: Hexagonal Blank

Speed of blanking, for one impression blanking tool, may be defined as number of strokes per minute. This generally depends upon the range of strokes per minute of power press. It may range from 30 to 58 strokes per minute or even more. Sturdiness of tool and sheet feeding mechanism may also be factors on which selection of speed depends.

Multi-impression tools are designed in such a way that generation of scrap (leftover) is minimum. Figure 2.8 illustrates layout of multi-impression blanking tool for a phosphor bronze blank.

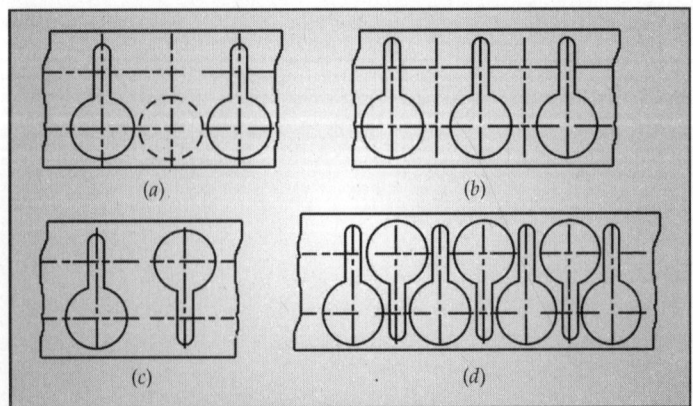

Fig. 2.8: Blanking Layout

Sheet blank as per dimension given in Fig. 2.9. A tie of 1.2 mm may be considered for the purpose of calculating percentage of scrap. Width of strip would be

$$\text{Length of blank} + 2 \times \text{tie}$$
$$W = 22 + 2 \times 1.2$$
$$= 24.4$$

As per layout shown in Fig. 2.8 (a), pitch of blanks (openings) would be 26.4 mm. So, in a strip of one metre length, number of blanks obtainable would be

$$\text{No of blanks} = 37.8, \text{ say } 37$$

Further, as per layout shown in Fig. 2.8 (b), pitch of two blanks would be 14.4 mm. So, in a strip of one metre length, number of blanks obtainable would be

$$\text{No of blanks} = 69.4, \text{ say } 69$$

Approximate area of one blank is calculated in following manner.

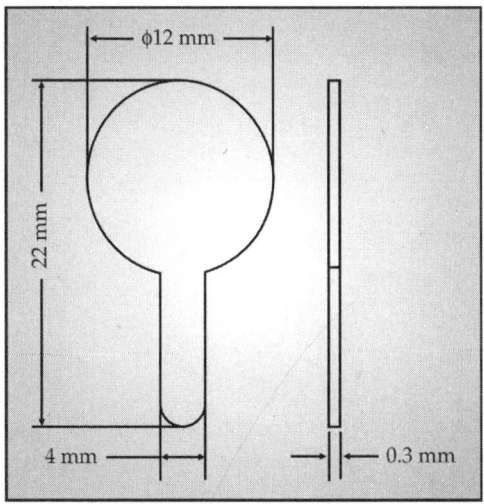

Fig. 2.9: Blank Dimensions

$$A_1 = 113 \text{ sq. mm}$$
$$A_2 = 32 \text{ sq. mm}$$
$$A_3 = 6.28 \text{ sq. mm}$$

Total Area = 151 sq. mm
Total Area of 37 blanks = 55.87 sq. cm

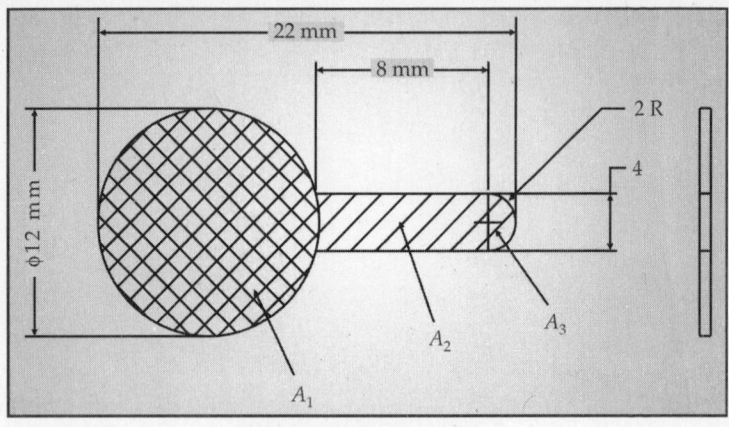

Fig. 2.10: Blank Area

Area of 1 metre long and 24.4 mm wide strip = 244 cm^2

Area of scrap	=	244 – 55.87
	=	188.13 cm^2
Percentage of scrap	=	77
Area of 54 blanks	=	81.5 cm^2
Area of scrap	=	162 cm^2
Percentage of scrap	=	66

From above calculation, it is clear that layout of blanking die (opening where punch precisely gets in) as shown in Fig. 2.8 (*c*) may be preferable as it generates less scrap as compared to layout shown in Fig. 2.8 (*a*).

Another typical component of mild steel sheet is shown in Fig. 2.11.

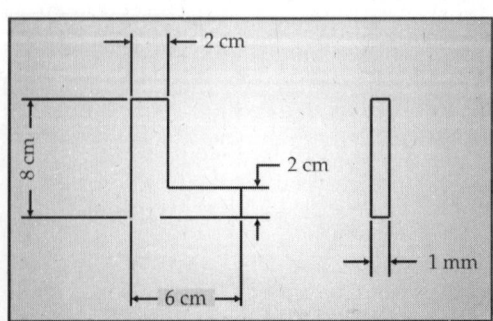

Fig. 2.11: L-Shaped Blank

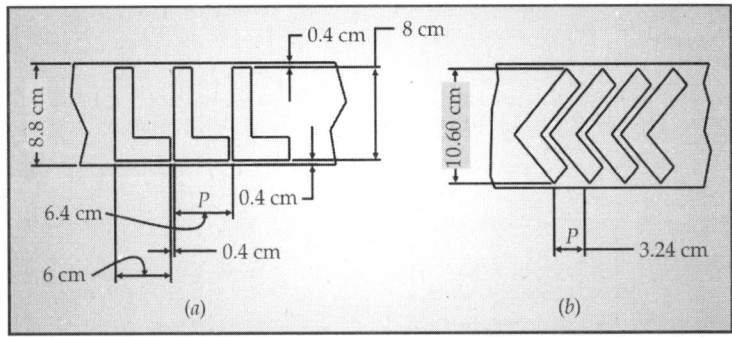

Fig. 2.12: Layout Possibility

Two possibilities of layout are considered, as shown in Fig. 2.12. In case of Fig. 2.12 (*a*), number of blanks which can be obtained in one metre length of strip work out to be 15 and for Fig. 2.12 (*b*), number of blanks work out to be 30.

Area of one blank = $8 \times 2 + 4 \times 2$
$$= 24 \text{ cm}^2$$
Area of 15 blanks = 360 cm^2
Area of one metre long and 10.6 cm wide strip = 1060 cm^2
Area of one metre long and 8.8 cm wide strip = 880 cm^2
Hence
Scrap area for layout as per Fig. 2.12 (*a*) = $880 - 360$
$$= 520 \text{ cm}^2$$
Percentage of scrap = 59
Scrap area for layout as per Fig. 2.12 (*b*) = $1060 - 720$
$$= 340 \text{ cm}^2$$
Percentage of scrap = 32

Above two examples demonstrate that percentage of scrap can be minimized by considering various possibilities of layout and calculating percentage of scrap. A layout with minimum generation of scrap may not be practically possible to adopt. This may be due to tool making limitations. So far such blanking process is considered where blanks are pushed through die. There may be typical possibilities where blanks are not completely blanked out of strip or sheet metal in the form of roll. Blanks are kept attached to strip for further operation in progressive tool design. Blanking operation is often combined with other operation like cupping. Hence a tool which carries out both the operations is called 'cut and cup' tool.

Trimming

In press working of sheet metal, trimming is an operation by which excess sheet is removed by circular cutting or flat cutting like a blanking operation. Examples of both the operations are explained with the help of Fig. 2.13.

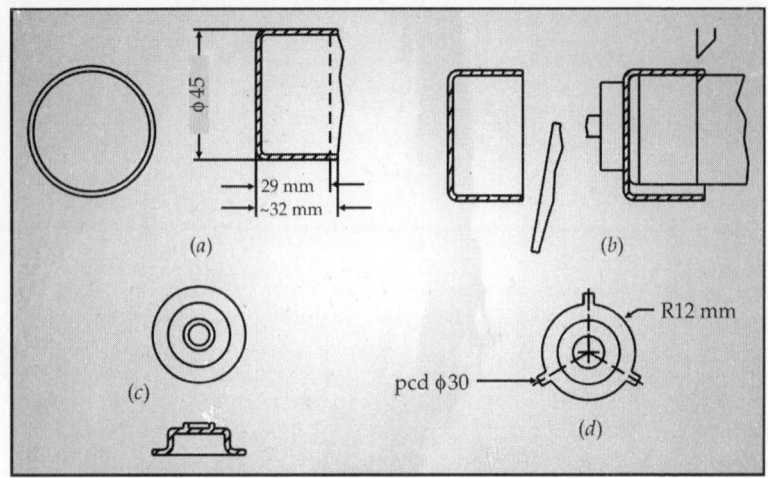

Fig. 2.13: Trimming Examples

Figure 2.13 (*a*) shows an aluminium cup which is produced by a 'cut and cup' tool. Brim of such cups is not perfectly straight. It needs to be straightened by trimming and to achieve specified height. In this typical example approximate height of untrimmed cup is 32 mm. Required height of cup is 29 mm. Figure 2.13 (*b*) shows a rotating circular trimming tool which cuts out a scrap ring to give a cup of 29 mm uniform height.

Figure 2.13 (*c*) shows another shallow cup component which has a flang of about 35 mm in diameter, which is not exactly circular. Figure 2.13 (*d*) shows the shape of desired flange which is about 5 mm less than roughly circular flange as shown in Fig. 2.13 (*c*). Final shape of brim of component (as shown in Fig. 2.13 (*d*)) is achieved by a punch and die of suitably matched dimensions and design. This operation with the help of tool is called trimming

As a general rule, die dimension produces blank dimension and punch dimension produces hole dimension. This general rule is taken into account if desired dimensions have very close tolerance.

Let the diameter of a blanking punch be 30.00 mm and the bore of die, be 30.03 mm. Then the hole generated by the tool would be 30.00 mm and the diameter of blanked disc would be 30.03 mm (Fig. 2.14).

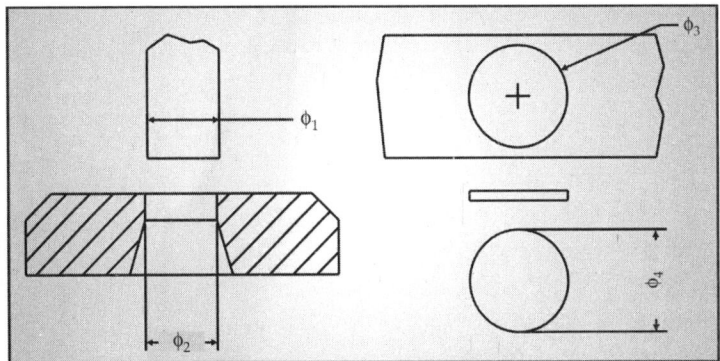

Fig. 2.14: Pierce and Blank

Bending

Bending is an operation by which portion of a strip or a component is made to have a different plain. It may be at a certain angle with respect to original plain. Figure 2.15 shows a few components before and after bending.

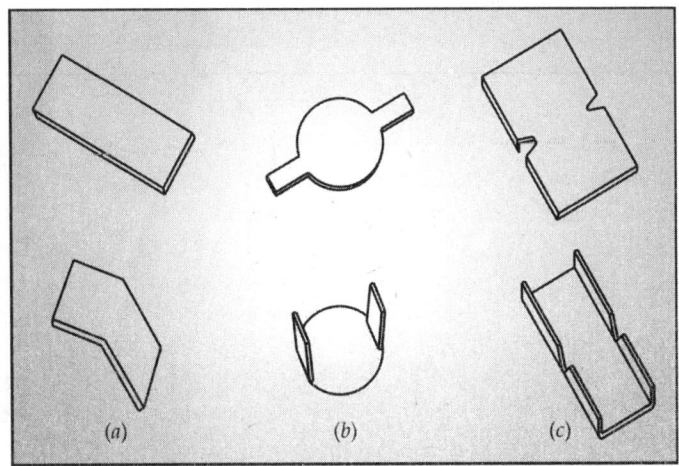

Fig. 2.15: Component Before and After Bending

Bending is a simple operation as long as bending angles or dimensions are not critical. Bending assumes a form of delicate operation if bending angles or dimensions become critical. Figure 2.16 is the drawing of brass strip before and after application of bending operation. In Fig. 2.16 (*b*), inner bending radius is 0.5 mm and in Fig. 2.16 (*c*), inner bending radius is 2 mm. Angle of bend in both cases is different. Bending is carried out by a tool shown in Fig. 2.17.

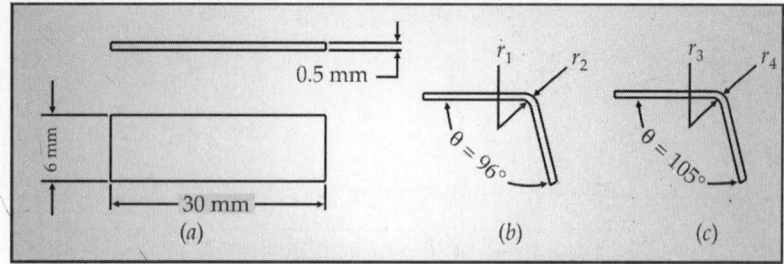

Fig. 2.16: Bending Operation

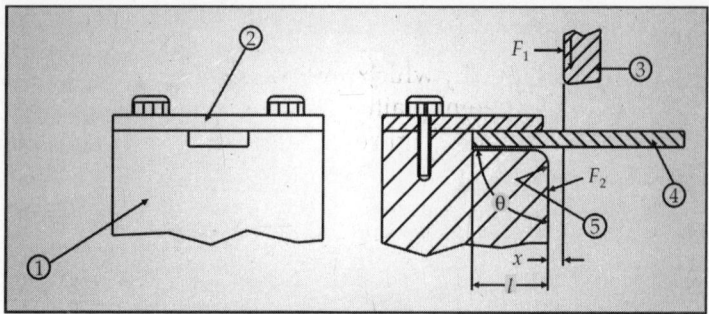

Fig. 2.17: Bending Tool

① is a die block which has a slot 19.5 mm long , 0.6 mm deep. ② is a cover plate. ③ is bending punch which is x mm away from die face. ④ is strip. In this typical example, x is kept as 0.55 mm. ⑤ is bending radius. When punch ③ travels down, it pushes the strip down which gets into the gap between punch and die faces F_1 and F_2. Face F_2 of die block is at right angle to slot face. After completion of bending operation, component is taken out of slot. On examining the component it is found that angle of bend is more than 90 degrees. This is because of 'spring back' characteristic of brass strip. How does spring back take place, is explained with the help of Fig. 2.18.

Figure 2.18 shows a magnified portion of bent strip. In the middle of thickness of strip an imaginary central line is drawn by dashes. On bending of strip, grains above central line tend to stretch. Maximum stretching of grains is near the surface of strip. Grains below the central line tend to compress. Maximum compression is near the opposite surface of strip. This stretching and compression of grains cause the tendency in the strip to 'go back' so that internal stretching and compression forces are reduced. Thus bent portion goes back a little after bending punch is pulled up. This phenomenon is called 'spring back'.

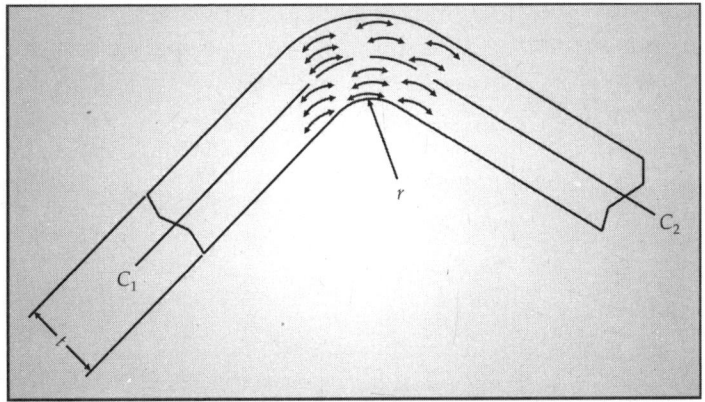

Fig. 2.18: Spring Back Reason

Amount of spring back depends on following factors.

- Thickness of strip
- Width of strip
- Hardness of strip material
- Grain direction
- Radius of bend
- Amount of extra bending
- Gap between punch and die working face.

Sheets of brass and phosphor bronze are produced with varying hardness. For example, an annealed phosphor bronze sheet may have a hardness of 100 BHN (Brinell hardness number) and the one having spring hardness of 119 BHN. Brass sheets may be 'quarter hard', 'half hard', 'spring hard' and 'extra hard'. Strip of quarter hard brass will spring back less after bending as compared to spring hard strip.

There are formulae available by which spring back of a particular strip may be calculated before designing and making a bending tool. Due to practical limitations, calculated spring back may not be exactly achieved. This may be, for example, due to the fact that strip actually does not have that Brinell hardness number which is taken for calculation. Bending radius may not be exactly what is taken for calculation. Grain direction in strip is not that which is assumed for the purpose of calculation. Gap between bending punch and die may be more than assumed for calculation. Slot in which strip is placed may be too high for strip to be too loose.

All above reasonings do not mean that calculations for spring back is of no use. No, it is useful to calculate spring back before finishing the tool. Calculation of spring back provides designer of tool with the nearest bending angles to be taken into consideration. Some fine adjustments may be done afterwards by actual bending trials. Following paragraph provides some information regarding calculation of bending spring back.

Parameters,
- θ_1 = Angle of bend in component
- θ_2 = Angle of bend in the die
- θ_3 = Spring back, (θ_2 minus θ_1)
- R_1 = Bending radius (internal)
- T = Thickness of strip
- M = Material of strip
- BHN = Brinell hardness number
- Grain direction

In case of half hard brass strip, spring back is usually around 8 degrees for a strip of 10 mm width, 0.4 mm thick and grain direction along the length of strip and bending angle as 45 degrees. Spring back would be more if brass strip is spring hard. If die and punch angle is 45 degrees and radius of punch (r_1) is 0.4 mm, spring back of 10 to 13 degrees may be expected.

Basic design of bending tool illustrated so far cannot take care of spring back if angle of bent in the component is 90 degree or more. A modified design of bending tool is shown in Fig. 2.19 which can be adjusted for achieving desired angle of bent in component.

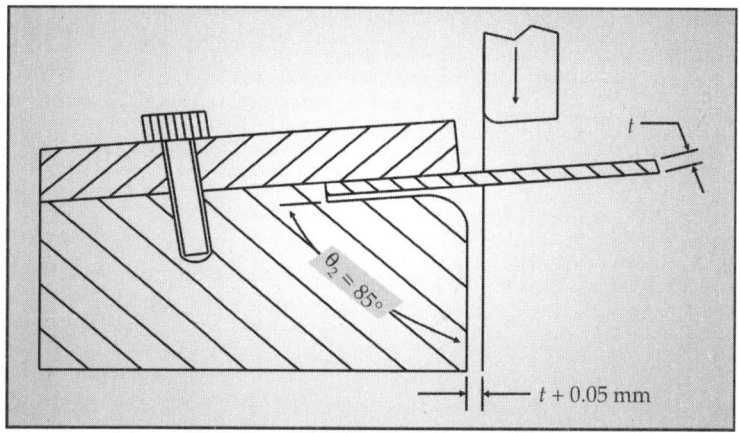

Fig. 2.19: Improved Bending Tool

In case bent strip is in V shape then basic design of bending die may be the one as shown in Fig. 2.20.

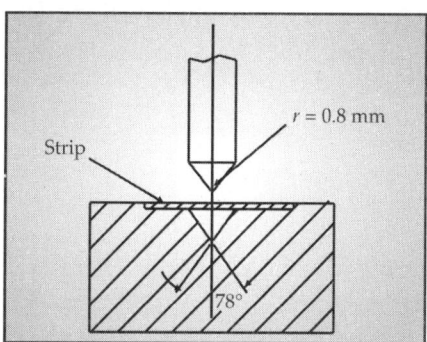

Fig. 2.20: V-Shaped Bending Tool

Some strips have got a number of bends as shown in Fig. 2.21.

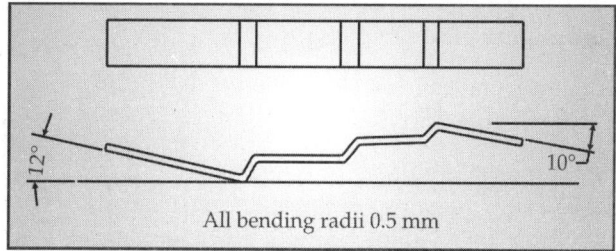

Fig. 2.21: Many Bends

Material of strip, as shown in Fig. 2.21, is phosphor bronze. All the bendings would be done at a time by tool of basic design as shown in Fig. 2.22.

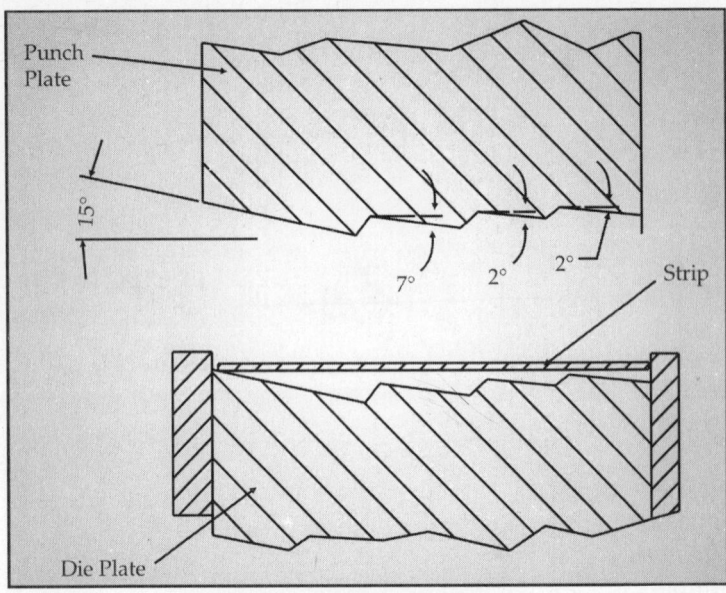

Fig. 2.22: Multi Bend Tool

In most probability one or two times correction in upper and lower tool may be necessary to achieve correct angles in component.

Stamping and Embossing

Stamping is an operation in press working of sheet metal by which an impression is created over the surface of a sheet metal part. Stampings are generally of two types:

- Fine and shallow
- Rough or fine and deep

In case of fine and shallow, only one-sided impression is created. Examples are fine marking on fountain pen metal clip, bottom of a pressure cooker, on metallic body of a pin stapler, etc. Figure 2.23 shows examples of both types of stamping.

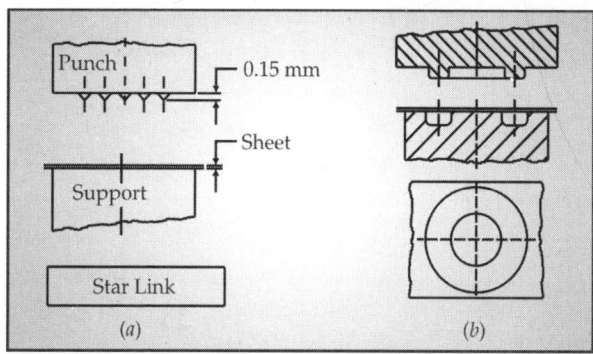

Fig. 2.23: Examples of Stamping

In Fig. 2.23 (*a*), a punch is shown which has raised letters. Height of such fine letters is kept about one-fourth the thickness of sheet metal. In Fig. 2.23 it is shown as 0.15 mm for a strip thickness of 0.6 mm. The other side of strip is completely flat.

In Fig. 2.23 (*b*), punch is shown which has raised ring die, has a sunken ring with increased dimensions to accommodate sheet thickness. On stamping impact, a raised ring is formed on sheet metal strip. This is illustrated in Fig. 2.24.

Instead of raised ring there may be any monogram or letter or numbers. In case of a coin, a blank of coin material is pressed/ stamped by two dies on a double acting press. Hence material flows in sunken details of die. Consequently, raised details develop on both sides of coin. This operation is called coining.

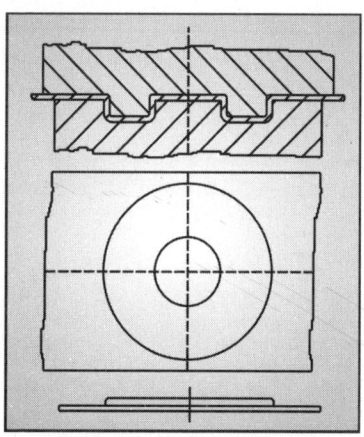

Fig. 2.24: Embossed Ring

Drawing

Drawing is an operation in press working of sheet metal by which flat sheet metal is partly or fully converted to the shape of a shell. Shell may be circular, rectangular or square. Figure 2.25 shows few typical cups, shell or container.

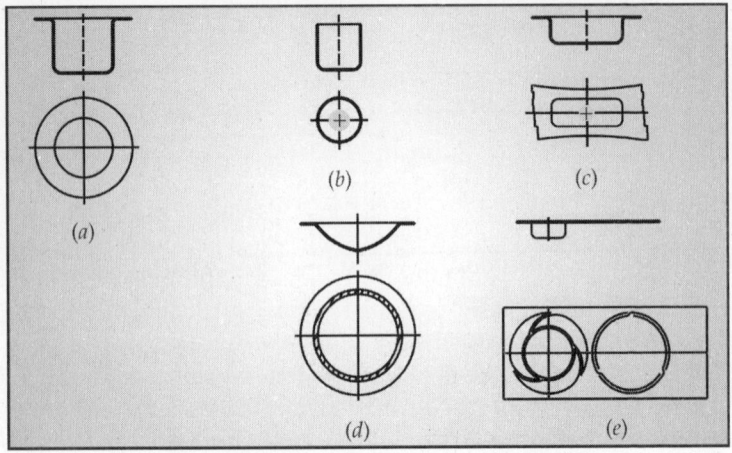

Fig. 2.25: Typical Cup/Shell/Container

Referring to Fig. 2.25, Fig. 2.25 (*a*) is a shallow round shell having a flange. Figure 2.25 (*b*) is a shallow round shell having no flange. Figure 2.25 (*c*) is a shallow rectangle container having flange. Figure 2.25 (*d*) is a dome drawn in a sheet metal. Figure 2.25 (*e*) shows a strip or coil of sheet metal to which drawn shells are attached by means of three curved ribs. Ribs are cut a station before drawing to hold a blank for drawing. This is to ease drawing operation.

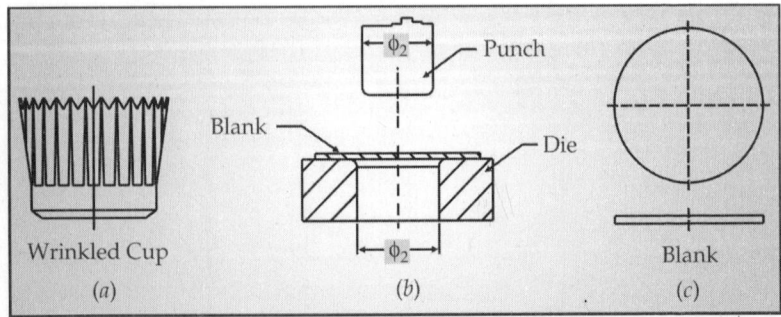

Fig. 2.26: Draw Punch and Die having Excessive Clearance

Figure 2.26 (*b*) is a brass sheet blank of say, 100 mm in diameter and 0.48 mm thick. Figure 2.26 (*d*) is die of 76 mm bore (ϕ). '*p*' is a draw punch of approximately 72 mm diameter, thus clearance between die and punch is 4 mm. Blank is placed over the die face and draw punch is brought down. It keeps on moving down till blank is completely cramped and gets into the die. Cramped cup (cup with wrinkles) is taken out which looks like as shown in Fig. 2.26. Formation of wrinkles is due to the fact that circumference of blank is forced to get reduced from $\pi \times 100$ to $\pi \times 76$ mm. In this typical example, circumference of blank cannot get reduced, therefore, it takes form of wrinkles so that it virtually gets reduced. Another situation may be that diameter ϕ_2 of draw punch is bore of die minus twice thickness of blank minus 0.05 mm. If draw is attempted resulting component would be something like shown in Fig. 2.27.

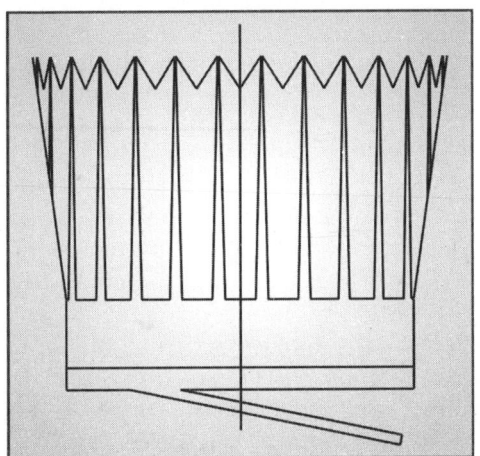

Fig. 2.27: Ruptured due to Wrinkles

For some distance, say about 3 mm, walls of cup will be straight and almost with no wrinkles. On further movement of punch, wrinkles would form and, therefore, blank cannot enter further, if punch is forced down, bottom of wrinkled cup would snap.

This problem is overcome by keeping the blank pressed while drawing is taking place. This is further explained with the help of Fig. 2.28.

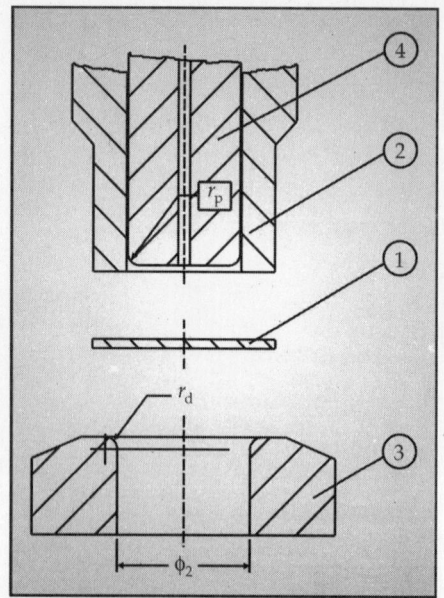

Fig. 2.28: Draw Tool with Pressure Pad

(1) is sheet metal blank. (2) is pressure pad which presses blank over the face of die (3), before the draw punch face touches the blank. Pressure between the faces of pressure pad and die face is kept so much that formation of wrinkles is avoided while punch is pushing the blank into the die. Draw punch and draw die have radii r_p and r_d respectively. Values of r_d significantly influences the draw quality of cup. For a particular blank, thickness and quality of material are the factors to decide value of r_d. As a general rule value of r_d may lie between 2.5t to 3.5t where t is the thickness of sheet.

Figure 2.29 (*a*) shows a circular blank which is marked by scratching or etching ink (Photo 2.6). Marked faces of blank are placed towards the face of draw die. After drawing operation, markings look like as shown in [Fig. 2.29 (*b*)]. Distance between two marked lines [(Fig. 2.29 (*b*)] is, say x. After cupping this distance become Δx. This means that grains of material in x distance are compressed in a distance Δx. Had there been wrinkles, no compression of grains would have been taken place. Due to compression of grains, hardness of cup wall is maximum at the brim of cup.

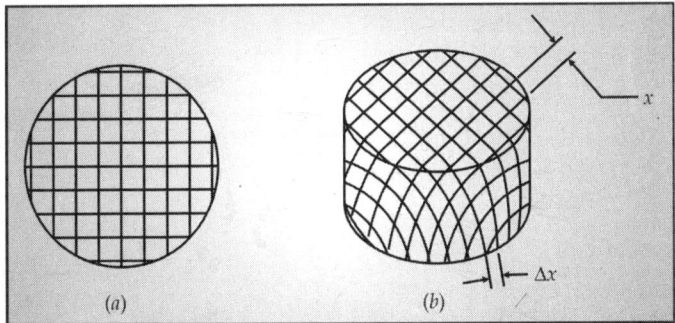

Fig. 2.29: Draw Pattern in Cupping

It can very well be appreciated that compression of material cannot be unlimited. In case diameter of blank is too large as compared to diameter of cup, complete cupping cannot take place. Force of drawing punch would rupture the base (closed end) of cup.

In practice, maximum percentage of reduction in diameter is about 40%. Beyond this percentage it becomes difficult to control flawless drawing of cup. Sometimes internal stress, due to compression of grains, near the brim of a brass cup, become so much that 'season cracking' takes place. This is very serious defect. To avoid this situation, stress relieving operation is carried out. It is done by placing cups/components in an oven at a particular temperature and for predetermined period of time. After stress relieving some finishing operations like 'bright dipping' may be necessary.

Deep Drawing

Deep drawing is a series of drawing operations where reduction of diameter from blank to diameter of final component is much beyond 40%. Hence, deep drawing operation cannot be carried out in one draw. A typical example of deep drawing is described with the help of Fig. 2.30.

A shell of 30 mm diameter and approximately 112 mm long is to be drawn from a circular blank of 120 mm diameter. This means a reduction of 90 mm in diameter. Percentage of reduction works out to be 75. This is just not possible. As mentioned earlier, reduction more than 40% offers difficulties. Therefore, practically it is tried to keep percentage of reduction less than forty.

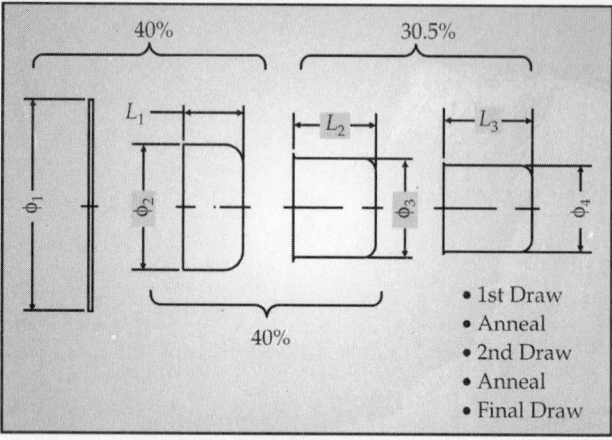

Fig. 2.30: Deep Drawing Stages

S. No.	Description	Diameter mm	Reduction in diameter	%age of reduction
	Table 2.1			
1	Blank	120	–	–
2	1st cup	72	48	40
3	2nd cup	43.2	28.8	40
4	3rd cup, final	30	13.2	30.5

Table 2.1 shows planned percentage of reduction. After first draw from blank to cup, grains are compressed maximum near the brim. Consequent to this compression, brass becomes hard and stressed. For further drawing it becomes necessary to anneal the cups to remove excessive hardness. Annealing is carried out on most suitable, practically determined temperature and period of time. Generally, temperature of annealing furnace is kept 475 degree centigrade for approximately two hours. Before 2nd draw, first draw cups are washed and rinsed and then 2nd draw is carried out. Same process is carried out for third draw. Practically, temperature and time depend on shape of components, number of components in the basket and heating capacity of furnace.

Progressive drawing is the process which is carried out by a special press, progressive dies and transfer mechanism.

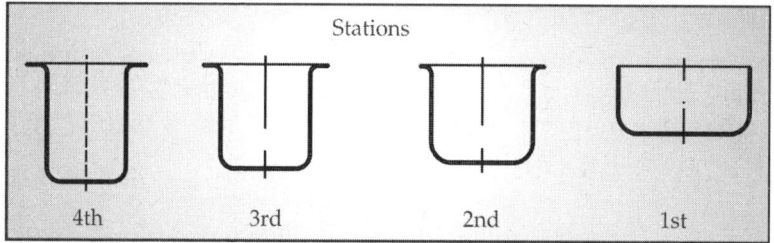

Fig. 2.31: Progressive Draw

Components are automatically transferred from one station to another. There is no intermediate annealing. This is due to the fact that grains are still in pliable condition.

Piercing

Piercing is a process by which holes are generated in a component. Multiple piercing is also carried out. Figure 2.32 shows a component which has three holes. These holes are made by a single stroke of punches.

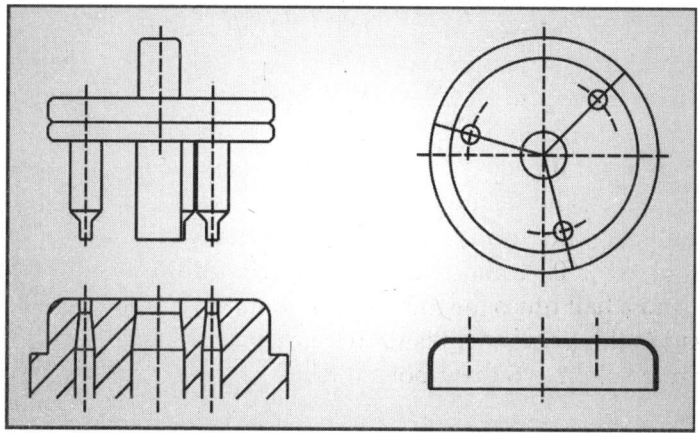

Fig. 2.32: Piercing Example

Size of pierced holes depends on punch diameter. Pierced holes may be non-circular also. Piercing punches and dies are designed and made accordingly.

Beading

Beading is a process in press working of sheet metal by which certain portion of a component is raised. Figure 2.33 shows a number of components in which bead formation is there. Beads may be in a flat surface as well as circular.

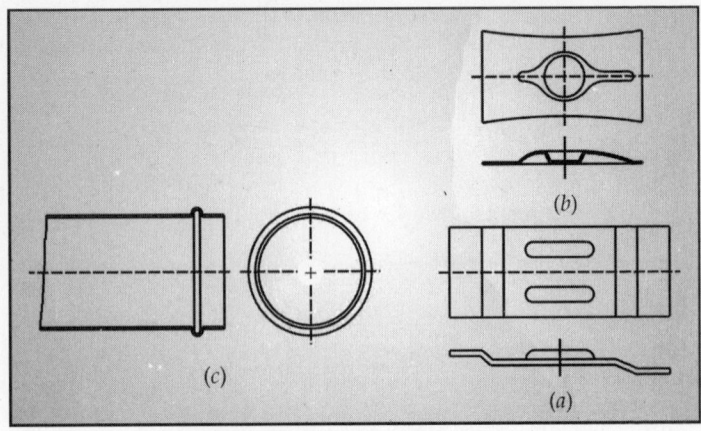

Fig. 2.33: Beading Example

Figure 2.33 (*a*) shows a bent strip component in which two beads are made near the edges of strip. Here the purpose of making beads is to strengthen strip against bending. As a general rule height of bead may be two to five times the thickness of sheet. Height of bead very much depends on type, thickness, hardness of sheet and width of bead. It needs a matched punch and die to form a bead. Figure 2.33 (*b*) shows another flat strip on which a shaped bead is formed. Height of bead is about four times the thickness of brass sheet which is 0.34 mm thick. Note that width of bead is about three and a half times the thickness of sheet. Purpose of formation of bead in this component is to strengthen aesthetic look and give a definite height to the component which is to be blanked out from strip.

Figure 2.33 (*c*) shows a circular component, say a tube of about 36 mm bore and 0.5 mm wall thickness. It has a bead of 40 mm diameter and 3 mm wide (outside dimension). Such types of beads are generally generated by mechanical means. Rotating tools are normally used. Basic working principle of rotating beading tools (die and punch) is explained with the help of Fig. 2.34.

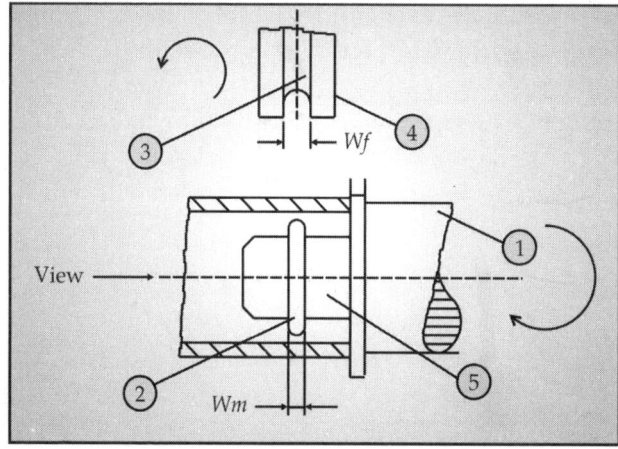

Fig. 2.34: Bead Rolling Tool

① is a spindle of a machine, rotating in clockwise direction when viewed in direction of arrow. This is a protruding portion ⑤ in spindle ①. ② is a bead turned and ground in spindle. Diameter of this steel bead is about four to five millimetres less than the bore of tube so that it may be inserted over lower beading roller and taken out after beading operation. Width of steel bead in lower roller is 1.5 mm having a semicircular shape as shown in Fig. 2.35.

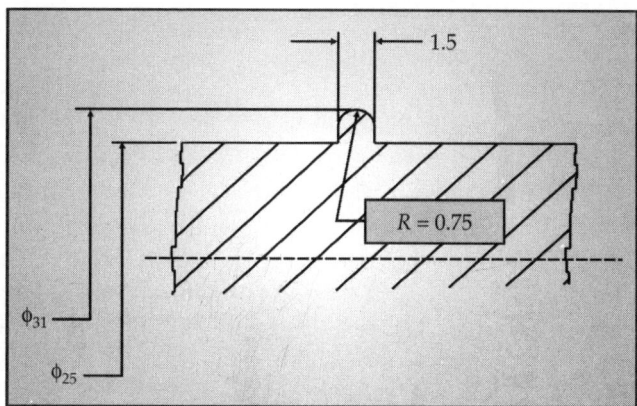

Fig. 2.35: Shape of Bead on Lower Tool

Design of beading machine is such that upper spindle can be lowered or raised with respect to lower spindle, yet without unwanted play in machine parts or sacrificing parallelism of upper and lower spindles. In Fig. 2.34, ④ is upper beading roller with a female bead formation ③. Width of this female bead is kept so much that $2 \times t$ ($2 \times$ thickness of tube wall) plus width of steel bead in lower roller are accommodated, this means

$$
\begin{aligned}
Wf &= t + Wm + t \\
&= 0.5 + 1.5 + 0.5 \\
&= 2.5
\end{aligned}
$$

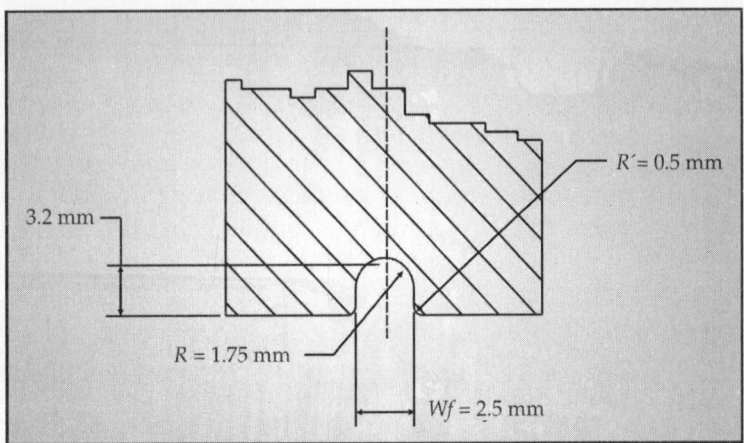

Fig. 2.36: Shape of Female Bead on Upper Tool

A small radius R is provided on the edges of female bead of upper roller. This is to facilitate flow of material during beading operation.

It is highly desirable that peripheral speed of lower roller and upper roller should be almost equal. Sometimes due to practical limitations due to machine construction, upper roller cannot be lowered to touch lower roller. In such a situation, designing of upper and lower roller is done in such a way that circumferential speed of lower and upper rollers is almost equal, whereas rpm of upper roller is exactly half the rpm of lower roller. Complete calculation is explained with the help of Fig. 2.37.

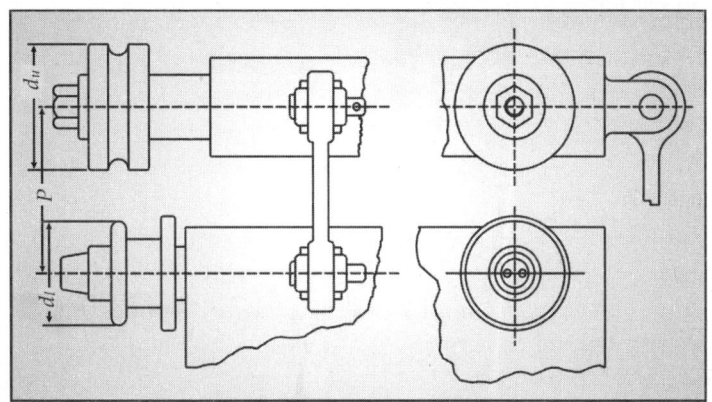

Fig. 2.37: Beading Setup

Let	p_{min}	=	minimum pitch	35 mm
	p_{max}	=	maximum pitch	65 mm
	d_u	=	diameter of upper roller	?
	d_1	=	diameter of lower roller	31 mm
	C_u	=	circumference of upper roller	?
	C_1	=	circumference of lower roller	97.4 mm
	N_u	=	rounds per minute (rpm) of upper roller	0.5
	N_1	=	round per minute (rpm) of lower roller	1

Circumferential speed of upper roller------- rpm × circumference

$$N_u \times C_u$$

Circumferential speed of lower roller------- rpm × circumference

$$N_1 \times C_1$$

Now
$$N_u \times C_u = N_1 \times C_1$$
$$0.5 \times C_u = 1 \times 97.4$$
$$C_u = 194.8 \text{ mm}$$

Therefore, d_u = 61.98, say 62 mm

Purpose of keeping circumferential speeds of upper and lower rollers approximately equal is to avoid rubbing of upper roller on component under beading operation. In case rubbing takes place, two unwanted things may happen. Firstly work hardening of component material at bead portion. Secondly, built up of material of component, say brass, on upper roller bead formation (steel bead). This needs frequent cleaning, causing stoppage of production.

Notching

Notching is an operation by which a small cut is produced at the brim of component. Notch can be made on flat portion of a sheet metal component or on the wall of round component. Figure 2.38 shows two typical examples of notches.

Generally purpose of notching is to locate component or product in a perticular place. Figure 2.38 (*a*) shows a pre-focus miniature lamp. A notch is cut on the collar of brass cap. This notch is extended a little in the body of cap. In Fig. 2.38 (*b*), a can is shown with a semi-circular notch.

Basic concept of notching tool (die and punch) is shown in Fig. 2.39.

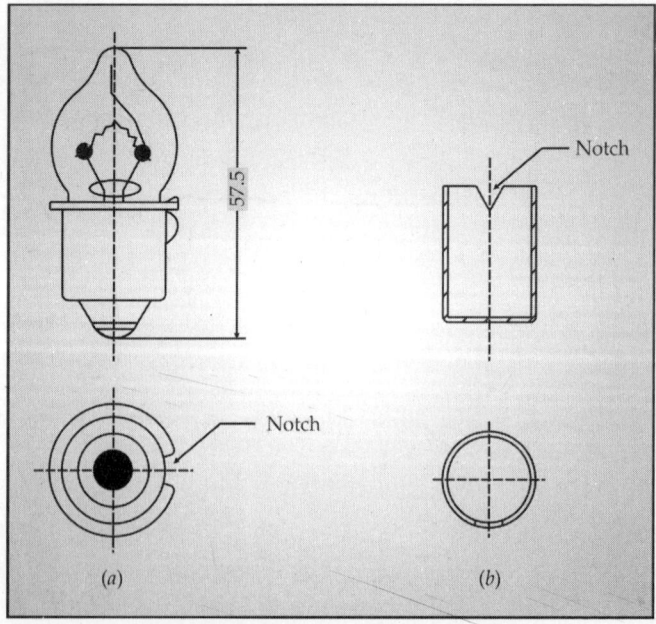

Fig. 2.38: Notching Examples

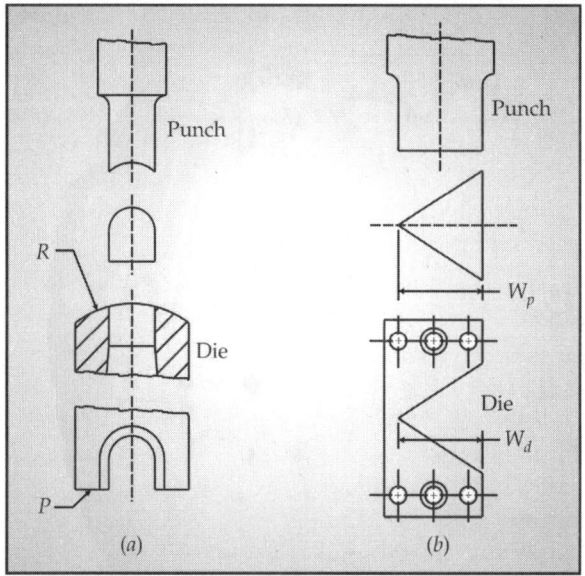

Fig. 2.39: Notching Die Concept

Figure 2.39 (*b*) shows a punch and die for generating a triangular notch. W_p is the width of punch and W_d is width of die which may be longer than W_p. Length of notch in a component depends on how much deep component brim is inserted into the die, but not more than W_p, width of punch.

Figure 2.39 (*a*) shows a notching tool, suitable for a circular component, say an aluminium sheet tube. Radius *r* of die is slightly less than radius of bore of tube so that portion to be notched definitely sits properly on the curved die. Depth of notch would depend on the depth of feed of component on the length of die. *P* is a point from where semicircular (may be of other shape) cut opening in the die rod starts. Suppose brim of tube crosses this point only by 1.5 mm then length of notch would be 1.5 mm. If tube crosses the point *P* by 3 mm, the length of notch would be 3 mm.

Thread Rolling

Many sheet metal components having rolled threads are used for general purpose and technical use. Normally purpose of threads in sheet metal components is to join parts, close opening or perform some function in a product assembly. A few examples of sheet metal thread rolled components are closures of a tin canister, assembly of 'ring' over the 'head' (where reflector is held) of a metallic, flashlight and base which acts as closure as well as focus adjuster. By rotating the base in clockwise or anticlockwise direction, focus of flashlight beam changes, this is an example of functional use of thread. Threads rolled in sheet metal components may be specified in following manner.

- Direction of spiral, that is, right hand or left hand (clockwise or anticlockwise)
- Single or multi start
- Pitch of threads
- Lead of threads
- Profile of thread on diameter
- Profile of thread in bore
- Depth of thread
- Shape of start of thread
- Shape of end of thread
- Thickness of sheet at threaded portion
- Matching of threads of two mating components

When a nut is turned in clockwise direction, it axially moves away from source of rotation, say hand of a person. Such a nut or bolt is called to have a 'right hand thread'. If nut is rotating in anticlockwise direction and moves away from source of rotation such a bolt or nut is called to have a 'left hand thread'. Same rule applies to threads on/in a sheet metal component.

Figure 2.40 (a) shows a sheet metal component which has a single start thread having pitch p. Figure 2.40 (b) shows a sheet metal component which has a double start thread having a pitch p and lead l as $2p$. In case of double start thread, starts of two threads are at 180 degree. In case of a three start thread, angle between two starts is 120 degree. Lead in a three start thread is equal to $3p$.

There are practical situation where space (length of thread portion) is not enough to generate more than one and a quarter thread. Under such circumstances grip between two mating parts is not strong enough. Two start threads overcome this problem to a great extent. Furthermore, there are practical situations where it is desired that quick screwing and unscrewing of two mating components take place. For such situations, two or three start threads are used. Figure 2.41 shows a number of typical thread profiles.

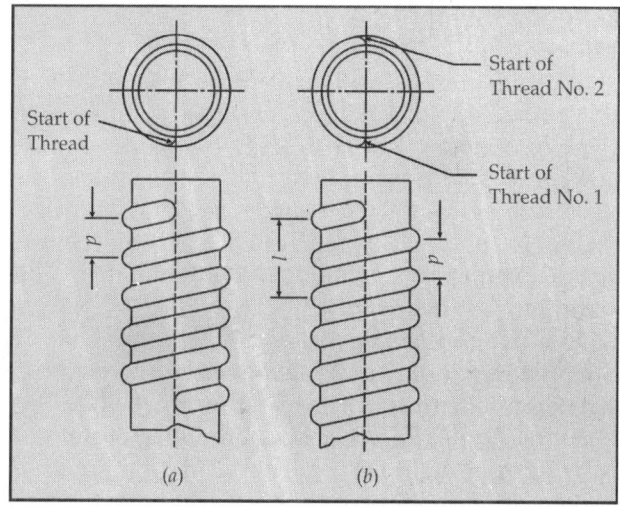

Fig. 2.40: Thread Start

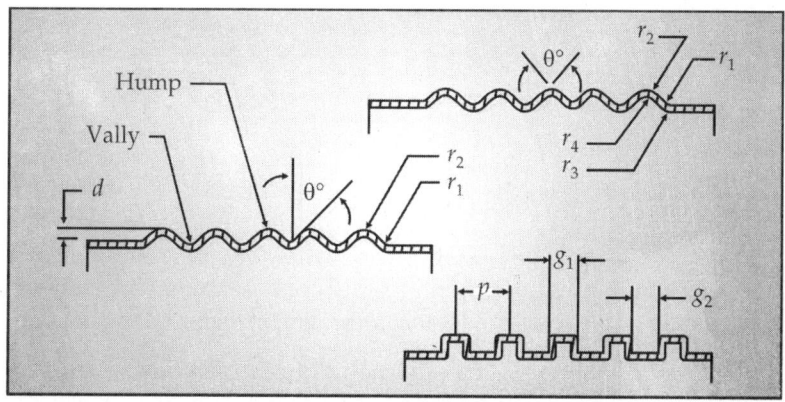

Fig. 2.41: Thread Profile

Figure 2.41 shows various parameters which are considered at the time of designing thread rolling rollers together with type and thickness of sheet metal. r_1, r_2, r_3 and r_4 are radii which are desired to be maintained. Sometimes, it is not practically possible to maintain very small desired radii. Grains of sheet get 'opened' with very sharp radii, say 0.3 mm for a sheet of 0.3 mm thickness. θ is acute angle of a 'hump' (raised portion of thread). This varies according to design of thread. Practically it may be 20 to 40 degrees. Acute angle of valleys (sunken portion of thread) would also be the same as that of hump. p is the pitch of thread. This means distance between the centre of two humps or valleys.

In case of multi start thread, pitch p is the distance between hump centre of one thread and hump centre of the other. But lead of thread is double the pitch in case of a double start thread. g_1 is approximate width of hump and g_2 is width of valley. In practice, it is quite difficult to measure width of hump or valley by means of a measuring instrument, say a vernier callipers. For proper measurement of depth of thread, radii, angles and width of hump and valley, it is necessary to draw a profile with the help of a profile projector. A typical thread profile is shown in Fig. 2.42.

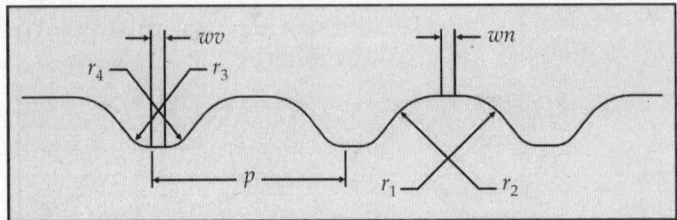

Fig. 2.42: Enlarged Thread Profile

All measurements can be directly read on profile projector with the help of a suitable built in scale image. Alternatively a shadowgraph can be drawn by placing a tracing paper on display screen. If profile projector is computerized, an attached printer can be set to get a photo printout. Thread shown in Fig. 2.43 is raised thread. Both raised and sunken threads are shown in Fig. 2.43. In case working or mating side of thread is inside the bore of sheet metal barrel/tube, taking of profile is a little difficult. Either a mould is to be lifted or a narrow strip is to be cut without damaging the profile. Sometimes, both internal and external sides of thread are working sides.

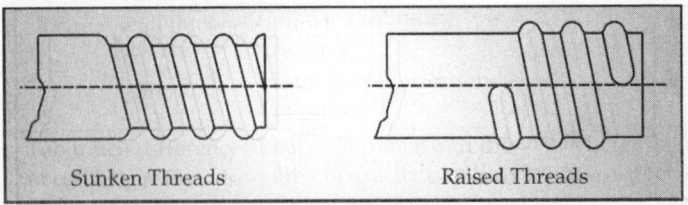

Sunken Threads   Raised Threads

Fig. 2.43: Sunken and Raised Threads

In Fig. 2.44, part 2 has right hand thread. In the base of this part, part 1 is screwed in and on the diameter, part 3 is screwed over it. In such circumstances, design of all the three components is critical, especially if the pitch of thread is to be kept fine. Fine thread means less space to form thread profile. Most difficult formation of thread would be in part 1 as width of hump of thread would have to be as minimum as possible so that it screws in easily in part 2. There would be hardly any difficulty in rolling threads on part 3.

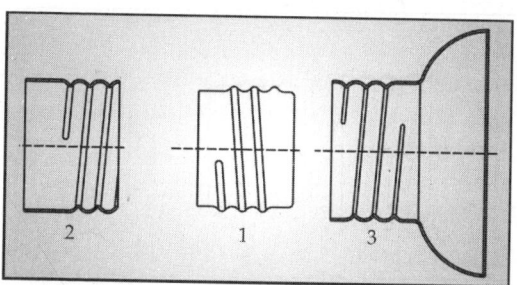

Fig. 2.44: Thread Pitch Consideration

Figure 2.45 shows a theoretical assembly of threads of three components.

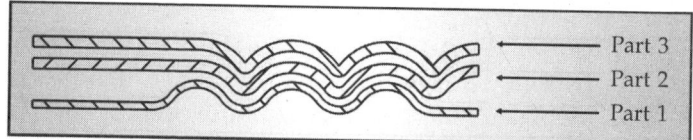

Fig. 2.45: Mating Thread Profile

Continuous threads are normally generated in sheet metal components. But there are instances where thread humps are intermittent. A typical component is shown in Fig. 2.46 which has intermittent thread.

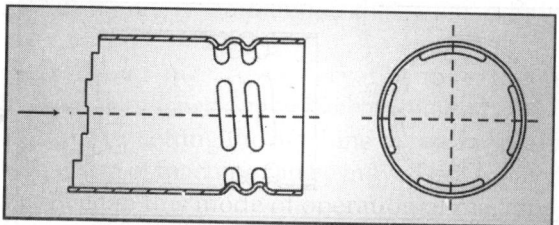

Fig. 2.46: Intermittent Threads

Intermittent thread design is adopted where diameter of component is so small that it becomes difficult to maintain circularity if continuous threads are generated. Furthermore, body of component remains strong and smooth if slided in bore of another component.

In this paragraph design of thread rolling machine, design of thread rollers, quality aspect and thread rolling process are discussed. Figure 2.47 shows fundamental construction of a thread rolling machine. ① is the upper cast iron body of machine which is fitted to main body ② in such a way that it can swing up and down with the help of reciprocating vertical arm ⑥ (Fig. 2.48). There is one precisely assembled rotating spindle ③. This spindle has no unwanted play as thrust bearings are used to eliminate unwanted axial play. At the same time there is such a mechanical arrangement that upper spindle can be very precisely shifted in axial direction with the help of a micrometer knob ⑦. Minimum axial displacement may be, say 0.02 mm by micrometer knob. Running of upper spindle diameter and roller support face ⑬ is highly true running. Run out is not more than 0.005 mm. Lower spindle ⑭ is also precisely assembled. Axial play is just enough for a lubricant film, say 0.02 mm. Lower spindle is hollow and has a Morse taper for fitting of an arbor which carries lower thread roller disc and other auxiliary discs. Upper spindle has a key so that upper roller and other discs do not slip rotate over spindle. Arbor has a locating pin protruding from the face. Lower thread roller disc and other discs have holes to suit protruding pin.

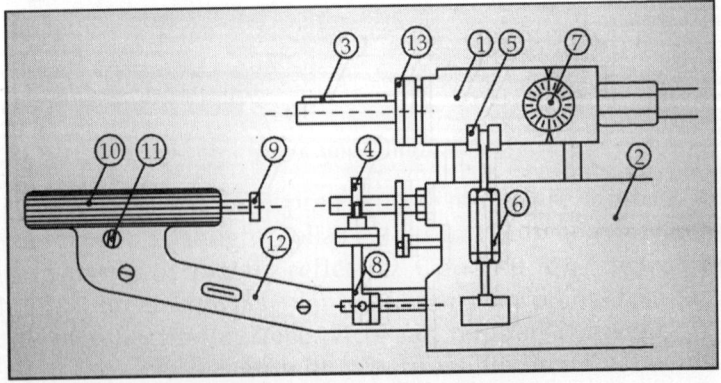

Fig. 2.47: Thread Rolling Machine

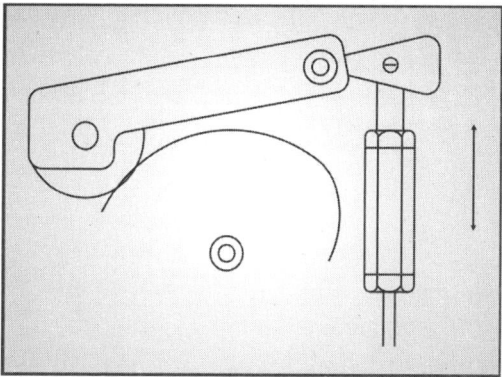

Fig. 2.48: Vertical Movement

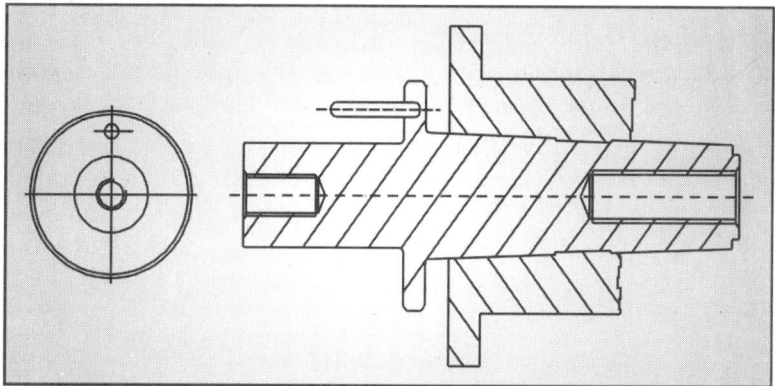

Fig. 2.49: Arbor

So there cannot be a relative motion between roller disc and arbor. Morse taper in lower spindle and arbor is so precisely machined that run out on arbor diameter and face is not more than 0.005 mm. Described accuracy is necessary for achieving good quality threads. ⑧ is an arm which swings to bring unthreaded component in front of lower spindle. ⑨ is back rest which pushes the component over lower thread roller and remains in that position till rolling is completed. It comes back just before ejector ring ⑭ starts moving out over lower tool to eject threaded component. Sequence of operation of automatic thread rolling machine is as follows.

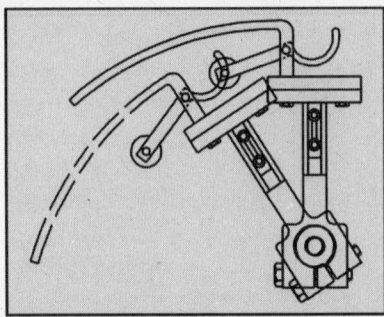

Fig. 2.50: Feed Arm Movement

- Manual or automatic feeding of component in chute of thread rolling machine
- Switching on the machine
- Both upper and lower spindles start rotating, lower in clockwise and upper in anticlockwise direction
- Back rest moves out and pushes the component over lower thread rolling tool. Stroke of back rest is adjusted according to need
- Upper body of machine moves down till threads of upper roller sink in threads of lower roller to such an extent that desired depth of threads in component is achieved
- Just before completion of thread rolling, feeder arm moves back
- On completion of thread rolling, back rest moves back and ejector ring moves forward to eject threaded component from over the lower thread rolling tool
- First few components are trial component when depth is adjusted by slightly rotating left-right hand long nut of vertical arm ⑥. Then profile of thread is checked to see if thread formation is symmetrical or there is need to do precision axial shifting of upper spindle.
- Machine is finally started to keep on producing threads on fed components

Upper and lower spindles are connected to each other by a train of gears towards the back end of spindles. By changing the gears combination, ratio of rotation of spindles can be changed. 1:1 means that both spindles rotate one complete round. 2:1 means that upper spindle makes one complete round when lower spindle make two complete rounds. Similarly a ratio of 3:1 is possible. These ratios are mentioned as example. It may be different according to machine design.

Facility to change spindle rotation ratio is very useful for designing upper thread rolling disc diameter. As mentioned earlier, peripheral speeds of thread rolling discs are kept as equal as possible. In following paragraphs a typical example is taken up to demonstrate as to how ratio of rotation of upper and lower spindles is decided.

Minimum adjustable distance between axes of
upper and lower spindles --- 40 mm
Diameter of lower thread rolling discs ---------------------- 28 mm
Diameter 'D' of upper thread disc should be more than ----- 52 mm

Say, 60 mm

This means that $N_l \times D_l = N_u \times D_u$
Hence
$$N_u = N_l \times D_l / D_u$$
$$= 1 \times 28 / 60$$
$$= 0.46$$

This means that upper spindle should make 0.46 of complete one round when lower spindle makes one round. Gears for setting such type of ratio is normally not possible. Nearest possible ratio which can be set by gear train is 1 : 0.5. This means that on one complete round of lower spindle, upper spindle would make complete half round.

If upper thread roller is made of 60 mm diameter then peripheral speed of upper thread roller would be more as compared to peripheral speed of lower thread roller.

Peripheral speed of lower thread roller at
600 rpm----- $= 600 \times \pi \times 28 / 1000$
$$= 52.8 \text{ metres/minute}$$
Peripheral speed of upper thread roller at
300 rpm----- $= 300 \times \pi \, 60 / 1000$
$$= 56.6 \text{ metres/minute}$$

Hence, there would be a rubbing action of 3.8 metres per minute.

This is an undesirable situation. Keeping the rotation of spindle as 1:0.5, diameter of upper thread roller disc would have to be recalculated.

$$N_u \times D_u = N_l \times D_l$$
$$0.5 \times D_u = 1 \times 28$$
$$D_u \quad = 56 \text{ mm}$$

Hence diameter of upper thread roller would have to be machined as 56 mm instead of 60 mm and lower thread rolling discs ---------------------- 14 + 28 = 42

This satisfies two important conditions:

- That peripheral speeds of upper and lower thread rollers are almost the same
- Total of two radii exceeds minimum distance between axis of upper and lower spindles.

Let required pitch of threads in the component be 2.5 mm. Hence the pitch of threads in lower thread rolling disc would also be machined as 2.5 mm so travel of a point will be 2.5 mm on one complete round of lower thread roller. But upper spindle has rotated only complete half round. This means that a point on thread of upper roller will travel only 1.25 mm (half of 2.5 mm). This is just not possible. Threads of both the rollers would severely strike each other and would get damaged. Solution to this problem is that two threads should be machined on upper thread roller. Start of both the threads should be 180 degree apart and lead of each thread should be 5 mm. Hence distance between the hump and valley of two threads would be 2.5 mm, which is pitch of threads. Figure 2.51 illustrates this explanation.

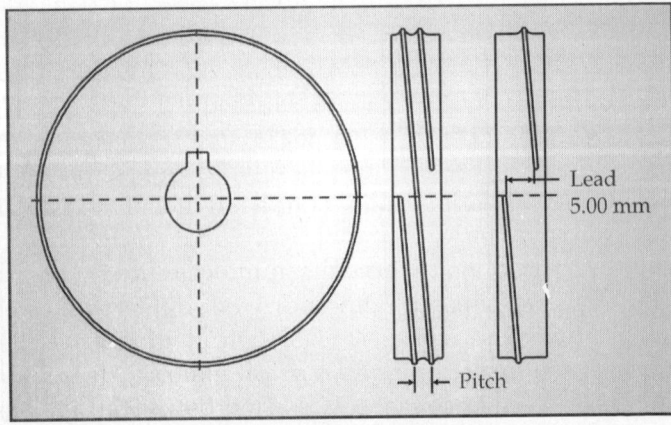

Fig. 2.51: Start of Threads

There may be cases where upper thread roller may need three start threads. Quality of threads formation very much depends on the quality of thread rollers and accuracy of rotation.

Figure 2.52 (*a*) shows correct matching of upper and lower thread roller's hump and valley. Symmetrical gap between hump and valley should remain unchanged if rollers are rotated. Also, vertical gap (*g*) should remain unchanged. Any roller running out on diameter would change vertical gap. It will be more at one place and less at 180 degree opposite to first place.

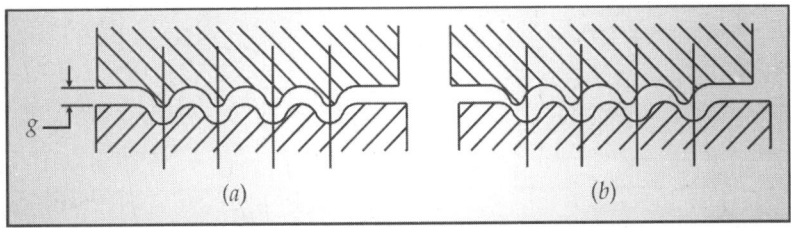

(*a*)   (*b*)

Fig. 2.52: Matching of Thread Rollers

Symmetry in axial gap depends on axial setting of upper spindle and true running of helix of threads. It is already mentioned that true running of diameters and faces of spindles is of extreme importance.

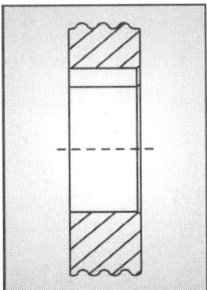

Fig. 2.53: Lower Thread Roller

Diameter, bore, threads and one face are normally machined in one setting on lathe or profile grinder. Before unloading the job from the chuck, a chamfer is made to indicate that this side of face is true to bore and helix of the threads. While assembling thread rollers on thread rolling machine, chamfer side face should be towards spindle collar face. In this way true running of thread helix may be ensured.

In practice, some additional discs are required to support thread rolling operation to achieve correct shape of threaded barrel/tube.

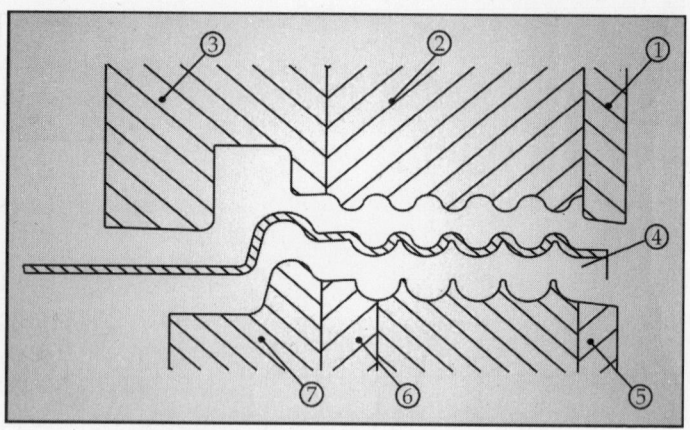

Fig. 2.54: Complete Thread Rolling Tool

In Fig. 2.54, ① is a plain rolling disc having a taper on diameter according to design of component ④. Normally taper angle is between 5 to 15 degree. Smaller diameter of taper disc is kept equal to diameter of thread roller. On many occasions it becomes necessary to slightly modify taper to achieve best results. This is very important that thickness of disc is equal on all points, say within 0.01 mm. Similarly thread rollers ② and ⑥, beading rollers ③ and ⑦ should have parallel faces very accurately. These accuracies ensure generation of good quality threads on sheet metal components. Inaccurate rotation of thread rollers and other discs cause unequal depths of threads or squeezing of sheet material, causing pin holes or cracks in threads.

Photographs

Photo 2.1

This is a vertical crank type power press of 40 tons capacity. A two impression cut and cup tool is loaded. Automatic roll feed system is attached to press bed. It has pneumatic clutch-cum-brake system. Pneumatic pipe and FRL unit can be seen. On left hand, lower side near the bed is fitted switching system. There are three push buttons to control ram movement. Push button on right hand is meant for inching. Push button on left hand is meant for running the machine in continuous mode. Big red push button in the middle is meant for emergency stopping of the press (*Courtesy: SISL*).

Photo 2.2

This is the close up view of two impression cut and cup tool which is loaded on power press (Photo 2.1). Two high pressure air pipes can be seen. These pipes provide high pressure air to create high pressure air jet to remove drawn cups from the tool and throw them on collection chute in the back of press. A four pillar die set can be seen. There are four springs around pillars to facilitate removal of upper portion of tool when tool is out of press for maintenance (*Courtesy: SISL*).

Photo 2.3

This is a progressive operation press. Ten square section rams can be seen which are moved up and down with the help of eccentric cams. Near the bottom, draw punches, fitted to rams can be seen. Underneath these punches there are die blocks. Pair of long horizontal strips can be seen. These are transfer arms, which transfer components from one station to another. In this way progressive drawing takes place (*Courtesy: SISL*).

Photo 2.4

An automatic thread rolling machine is shown. A number of operations like beading and trimming can also be performed. Magazine, feeder, backrest, thread rolling-cum-beading upper and lower tools can also be seen. Depth adjusting vertical actuating arm is also visible (*Courtesy: SISL*)..

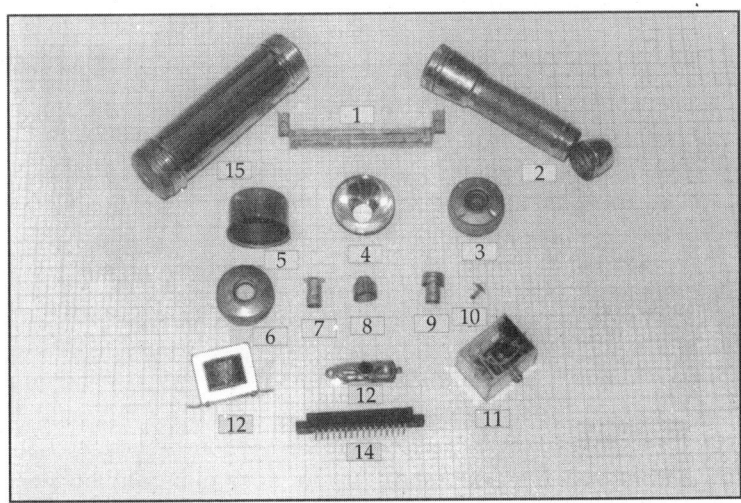

Photo 2.5

In this photograph a number of components are shown. Each component is affixed by number for easy reference.

Components

1. It is a clamp of sheet metal. A number of bends can be seen. There are two long beads to give strength to long portion of clamp.
2. A complete sheet metal flashlight is shown which has a number of sheet metal components. Closing cap with rolled threads and a contact spring is there. Circular ring formation on body is done by rolling tool.
3. Shows an assembly in which many sheet metal components are assembled. Only few components are visible.
4. It is a sheet metal reflector which is produced by stretch cupping, piercing- cum-trimming operations. Parabolic surface is vacuum metalised for good reflection of light.
5. It is a brass cup. Reduction from blank diameter to cup diameter is about 40%. This is quite a high percentage.
6. A sheet metal component which is produced by cut and cup, thread rolling, circular trimming, stamping and piercing operations.
7. It is produced by operations cut and cup, progressive draws, trimming, piercing and intermittent thread rolling.
8. This is another component of assembly ③. It is produced by cut and cup-cum-piercing and making of a notch which is a beading/forming operation.
9. This is sub-assembly of bulb holder (comp 3).
10. It is a small sheet metal component produced from spring hard brass sheet. Operations involved are multi impression blanking, forming and final bending.
11. An electrical relay has got a number of sheet metal components, like spring hard strip carrying contact points, bent components to hold spring and armature.
12. It is a transformer which is housed in a rectangular sheet metal casing having legs for assembly in any gadget.
13. It is a switch of a metallic flashlight. It consists of a number of sheet metal components. Visible components are main body and slider. Main body is produced by applying operations, draw, trim, bend, pierce and slit (rectangular piercing).
14. This is a connector having plastic body and a number of contact pins, made of brass sheet by trimming, piercing and bending operations.
15. Shows body of a sheet metal flashlight. Raised beads and rolled threads are visible. Lines on the body are produced by rolling embossing/ forming process.

Photo 2.1

Circular blank of this cup was marked by scratching the lines like a graph paper. It is then drawn. Pattern of draw can be clearly seen. At the brim of cup distance between the lines has become less (*Courtesy: SISL*)..

3

Impact Extrusion of Aluminium and Zinc

Underlying Principle

A variety of components in the form of containers or tubes can be found in market place. What is seen on retail shop counters are finished products where extruded containers or tubes were starting input in producing finished product. Construction of a dry cell, for example, is around a zinc can. Inside the can are various chemicals, components and carbon rod. Outer surface of can is covered by paper or metallic jacket. In some brands of dry cells, base of zinc can is also covered by some other component, for example, a tin disc. Zinc can is produced by a process known as reverse impact extrusion.

Metal body of a dry cell holder is another example. Body may be of brass, tin or aluminium. Aluminium container/body of a two standard cell $(2 \times R\ 20)$ is about 15 cm in length. Such a long aluminium tube cannot be extruded easily with thin wall. First of all a can of about 4.5 cm diameter is extruded and then it is drawn to a predetermined length.

Figure 3.1 shows a small set up to perform an experiment to explain the underlying principle. In Fig. 3.1, ① is a small clear glass or plastic container or cup of about 3 cm in bore. A round disc of hard butter (taken out from refrigerator) is placed in the container. A round polished rod of about 2.5 cm diameter is placed over the butter disc in such a way that by visual judgement it is in center. Now, while observing near the bottom of container, press the rod over the butter disc. It will be found that butter rises in the gap between container's inner wall surface and rod. It demonstrates that if a ductile (flowable) material is pressed in a confined space, the material would flow out through available space.

Suppose same experiment is to be performed on lead disc then a steel container and push rod would be needed. For forcing the rod/punch down, a press would be needed to provide required force and speed.

Before discussing actual reverse impact extrusion of aluminium and zinc, it is advisable to review some **mechanical properties** of aluminium and zinc, which predominantly affect cold extrusion process. In cold extrusion process, aluminium or zinc is subjected to deformation, which needs aluminium of 99.7 to 99.9 % purity, which is most suitable for cold impact extrusion. Process of impact extrusion needs sudden application of force, which make aluminium to flow. It takes fraction of a second to complete the travel of extrusion punch from the instance of punch face touching calot (aluminium disc) to its completion of travel. Completion of travel means that not beyond a certain amount of impact, so much work on hardening of aluminium or zinc takes place that further flow of metal is not possible. Construction and working principle of an extrusion press is explained with the help of Fig. 3.2

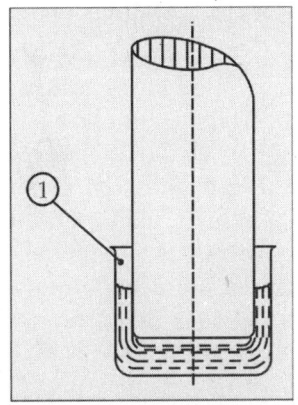

Fig. 3.1: Small Setup

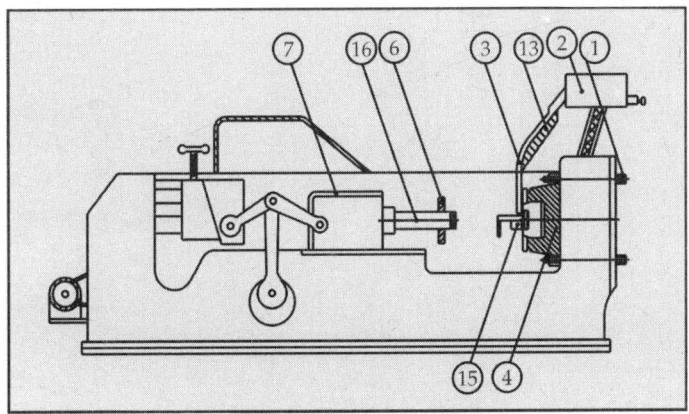

Fig. 3.2: Extrusion Press

Referring to Fig. 3.2, ① is a heavy duty cast steel body of a, say 110 tons impact capacity press. ② is a mechanical feeder in which lubricated aluminium calots are fed. Calots are tumbled with zinc stearate in tumbling barrel. Zinc stearates acts as lubricant, which improves the flow of grains. ③ is a chute through which calots are fed in front of extrusion die. Chute is fitted with electrical heaters ⑬ which keep the chute hot (say about 230 degree centigrade). Purpose of keeping the chute heated is to heat the calots to, say 180 degree centigrade till it reaches in the gripper of calot feeder ⑮. Movement of calot feeder is so synchronized with the movement of Ram ⑦ that feeder takes the calot exactly in front of extrusion die ④ just before face of extrusion punch reaches to push the calot inside the die for its extrusion. At this very moment, feeder looses its grip on calot. Extrusion punch ⑯ pushes the calot inside the die and extrudes it. Just after completion of extrusion in about one hundredth of a second, return stroke of punch starts. It comes out of die together with extruded can/container. Container/can is normally quite tight over the extrusion punch nose. There is an ejector ring ⑥ which is fixed to machine body ① in such a way that its axis and that of extrusion punch coincides. Design of ejector ring ⑥ is in the form of six segments, held together with a spring band. Further, the housing is so designed and constructed that ejector ring remains floated radially by about two to four millimetres. Ejector ring is stationery in axial direction except for clearance for floating action. Construction of a typical ejector is shown in Fig. 3.3

Normally there are six segments ② which are kept collapsed with the help of tension spring ①. In collapsed condition bore ϕ is almost circular and about 1.5 mm smaller than the diameter of extrusion punch.

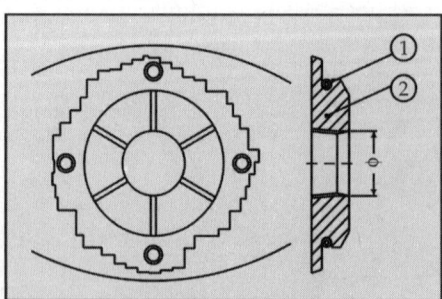

Fig. 3.3: Ejector

While extrusion punch is in return stroke with can, ejector stops axial movement of can. Hence can gets stripped from the punch

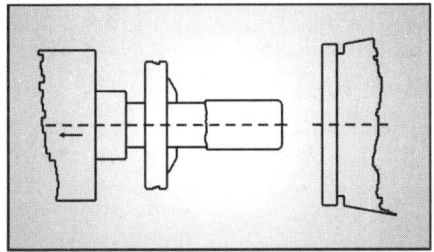

Fig. 3.4: Ejection Action

and fells down or transferred by a transfer mechanism on to the conveyor. Extrusion keep on repeating itself. Machine stops only if it is done so or a safety device gets activated due to some malfunctioning or opening of safety view gate. Basic designs of extrusion die and punch are explained with the help of Fig. 3.5

Let c be an aluminium calot of diameter ϕ of, say 38 mm and thickness t as 6 mm. Required diameter of extruded can is 38.30

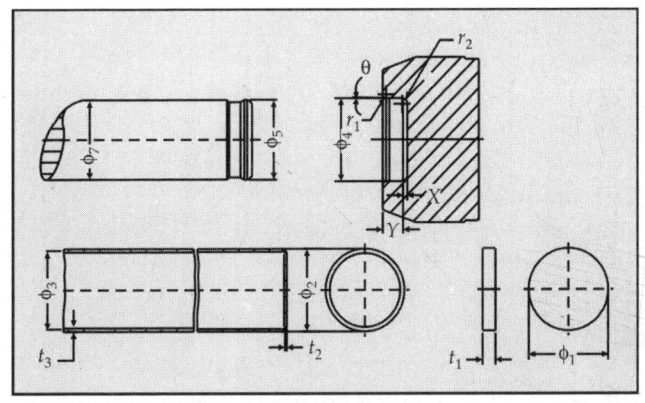

Fig. 3.5: Extrusion Tool

± 0.03 mm. Required bore of can is 36.30 ± 0.03 mm. Hence wall thickness t_3 of can would be

Minimum wall thickness ----- = Minimum diameter ϕ_2 – Maximum bore ϕ_3

$$= (32.27 - 36.32)/2$$
$$= 0.97 \text{ mm}$$

Variation in wall thickness should not be more than 0.05 mm. Required base thickness t_2 of can is 1 mm, ±0.03 mm. Wall and base of can should be reasonably free from scratches and scoring of metal. Cans should also be free from mechanical damages or distortion. Die is machined from steel, which has following dominating characteristics:

- Shock resistant
- High wear resistant
- Achieveable surface hardness during heat treatment, 60 Rockwell c scale
- Polishability

Suggested dimensions of die and punch for 'input calots' and extruded cans are given below:

X = Length of cylindrical portion of die. Normally it is equal to thickness of calot. In the example it is kept as 6 mm.

Y = Length of taper portion of die bore. It may be about double the thickness of calot, say 12 mm

θ = Angle of taper with respect to axis of die. It may be between 1 to 2 degree.

r_1 = May be about 2 mm

r_2 = It is a critical radius as flow pattern of aluminium very much depends on this. Theoretical determination of this radius can be done. But most suitable value is determined by some trial and error method. In this case it is about 4 mm.

Cold reverse extrusion punches made from shock resisting steel which can be heat treated to achieve a surface hardness of 58 to 59 Hrc. Steel should also have good wear resistant properties. Punch should be ground finished with specified 'nose' profile and dimensions.

Nose means some length of punch from working (extruding) face. Referring to Fig. 3.6, some dimensions are given and described as below:

ϕ_5 = Maximum diameter of punch nose. In the typical example it is 36.30 mm.

r_3 = Critical radius of about 3 mm

p = Length of nose covered by radius of 3 mm

l = Length of 'land', 2.5 mm

ϕ_6 = Relief diameter, 36.20 mm

ϕ_7 = Punch shank diameter, 36.28 mm

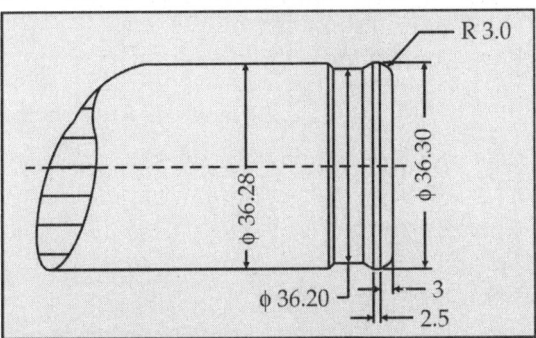

Fig. 3.6: Punch Nose

Combination of abovementioned dimensions of punch and die are expected to extrude good quality container/can. But abovementioned combination is not enough to achieve desired quality of can. Some more important conditions are to be fulfilled. At the time of stripping of can, a vacuum is generated between extrusion punch face and inner face of can. If thickness of can wall and base is not strong enough, it will collapse at the time of ejection. If it is strong, some amount of air leaks in from between punch nose diameter and bore of can to prevent collapsing. Sometimes dimensions and strength of can are such that collapsing of can is prevented only by providing an air vent valve. Construction of such a valve is shown in Fig. 3.7.

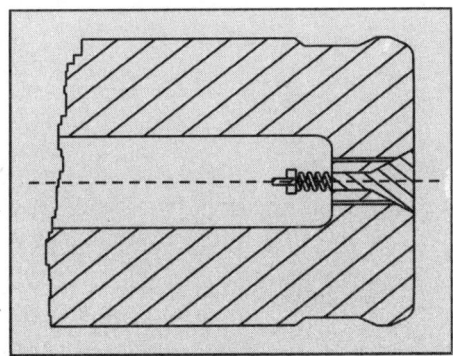

Fig. 3.7: Air Valve

Quality Aspects of Extruded Cans

For production of defect-free cans, following two aspects may be taken into consideration.

- Defining required quality norms.
- To have understanding of defects, causes and remedies

Normally desired quality norms are as follows.

- Cans should be mechanically strong enough to avoid denting/damage during normal handling and conveying
- Thickness of wall should be in specified limits at various points
- Base thickness should be in specified limits
- Length of can at various points should be under specified limits with trimming allowance
- Outer surface of can should be free from objectionable scratches
- Outer surface of can should be free from scoring
- Inner surface of can should also be free from scratches and scoring

Table 3.1: Some Insight into Defects, Causes and Suggested Remedies

S.No.	Defects	Causes	Remedies
1.	Can diameter more or less than specified	Incorrect bore of extrusion die	Replace die with the one which has correct bore
2.	Wall thickness more or less than specified	Dimension of punch nose incorrect	Replace punch with correct diameter
3.	Scratches on diameter of can	Scratches are there in the die	Polish die bore or replace it
4.	Scoring on base of can	Aluminium or zinc built up on the impact face of die	Polish out metal built up
5.	Scratches in the bore of can	Presence of scratches on punch nose	Polish punch nose or replace punch with a correct punch
6.	Unequal wall thickness	Punch and die are not coaxial. Improper lubrication of calots	Reset the die to make it coaxial. Use correctly lubricated calots
7.	Too much difference in can height at different points	Faces of punch and die are not exactly parallel	Check ram alignment with respect to die seat on machine body.

Table 3.1: Some Insight into Defects, Causes and Suggested Remedies *(Contd...)*

S.No.	Defects	Causes	Remedies
			If error is there, correct it. Check parallelism of die face and its back face which rests on machine body. Retool it if found incorrect.
8.	Waviness develops on walls of can during stripping from punch	Diameter of punch shank slightly more than punch nose diameter	Replace punch with a correctly dimensioned punch
		Length of 'land' more than specified	– do –
		Wall thickness of can is not strong enough to sustain stripping	Reconsider design dimensions of can
		Too much variation in can height at various places	Reduce variation by resetting the die

Safety Precautions

Manufacturer or supplier of extrusion presses normally supplies operating instruction manual. It should be studied and safety precautions should be followed. For example, clear plastic thick sheet window should always be in closed position before machine is switched on. There is a possibility of extrusion punch breaking up and debris flying out. These projectiles may be so powerful that it may cause fatal injury to setter, operator or an observer of machine. This may happen due to, for example, malfunctioning of feeder system and feeder finger of mild steel or carbon steel stays in front of extrusion die. Due to excessive loading, punch breaks. That is why it is highly recommended that safety window should be closed.

Sometimes it becomes necessary to move the punch for observation, alignment checking or synchronization of calot feeder. This may be done by setting the machine to run in 'inching' mode. This means that ram of machine can be moved gradually by manual push button. Even in this mode of operation of machine, operator should take care of self-safety and those nearby.

Extrusion Press and Tool Setting

Reverting back to Fig. 3.2, basic construction of a horizontal extrusion press is shown. As mentioned earlier, body of press is sturdy and made of cast steel special alloy. Ram of machine is quite heavy and sturdy to provide sufficient inertial force at the time of start of reverse cold impact extrusion of aluminium or zinc calot.

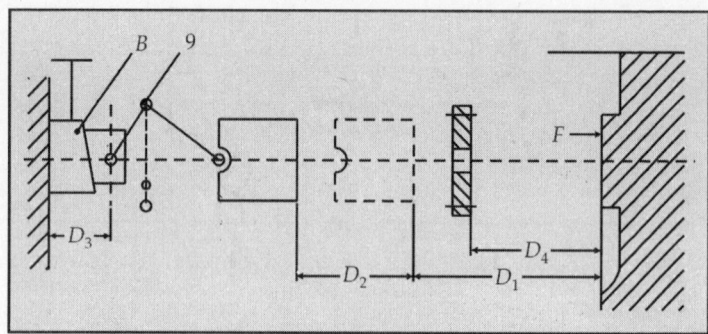

Fig. 3.8: Ram Position

An electric motor drives a heavy flywheel with an inbuilt pneumatic clutch. Drive from electric motor to clutch is through a speed variator. Rounds per minute of flywheel can be varied, say from 90 to 110. Pneumatic clutch output directly drives a heavy duty crank shaft which in turn drives a knuckle joint ⑨ as shown in Fig. 3.8. Length of ram stroke may be around 18 cm. In forwardmost position of ram, there is a definite distance between ram face and machine body face where die is fitted. Let it be D_1. D_2 is the distance of travel of ram. This is a fixed travel. Design and construction of extrusion press is done in such a manner that distance D_1 is sufficient to accommodate die holder, die, punch and punch holder. Travel D_2 of ram is so designed that distance between body face Ⓕ and ejector plate is sufficient to provide space for die holder, feeder system, transfer system and free space to do setting, etc. and free fall of extruded can.

Figure 3.9 shows the gap Ⓧ between faces of punch and die when ram is at its outermost position. This gap Ⓧ needs to be precisely adjusted to achieve desired thickness of can base. Facility of this adjustment is provided by a taper block Ⓑ which can be precisely moved up and down, thus creating change in distance D_3 which in turn changes value of Ⓧ.

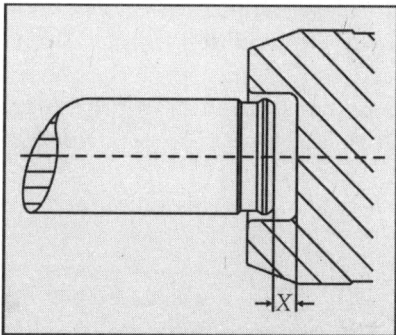

Fig. 3.9: Gap between Punch and Die Faces

As mentioned earlier, it is necessary to set die coaxial to punch. Die can be shifted by loosening and tightening the bolts of die housing. Arrangement of such adjustment is shown in Fig. 3.10

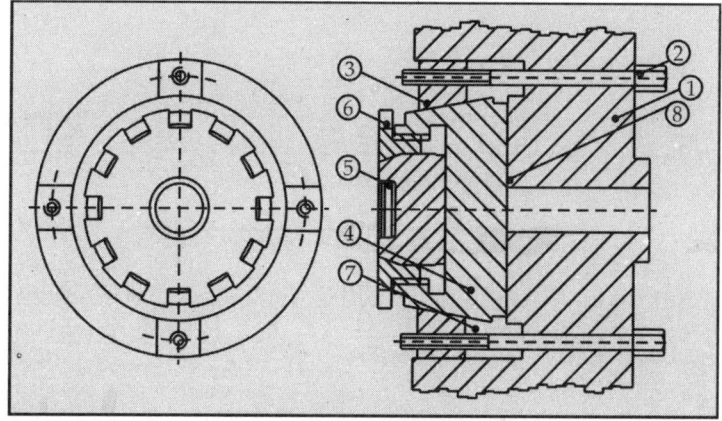

Fig. 3.10: Extrusion Die Fitting

Referring to Fig. 3.10, ① is body of extrusion press. ⑦ is a circular housing in the body of press. Face ⑧ and bore ⑦ of circular housing are very accurately machined surfaces. ② are four tightening bolts with specially shaped nuts ③. Die holder housing ④ is heat treated to about 56 HRc. Tapered diameter and faces are very accurately machined in one setting on cylindrical and surface grinder. ⑤ is extrusion die which is tightly held in housing ④ with the help of big nut ⑥ which has fine threads, suiting to housing ④. Suppose die is to be shifted downwards then lower bolt ② would be loosened a little and upper bolt tightened accordingly. But before doing this, two side bolts would have to be loosened a little to allow downwards shifting of die housing. After retightening all the bolts, trial extrusion would be done and walls of can are measured. In two or three attempts uniform wall thickness is achieved. This adjustment very much depends on practice of operator in setting and resetting of extrusion die.

Arrangement of holding extrusion punch is also shown in Fig. 3.11.

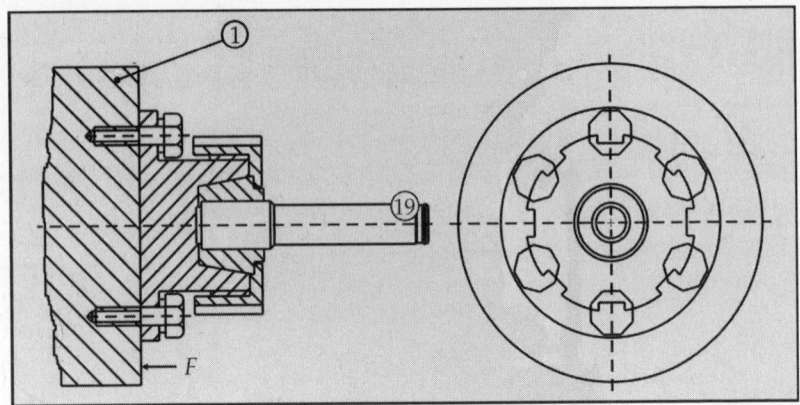

Fig. 3.11: Punch Holder

Referring to Fig. 3.11, ① is ram of extrusion press. Face Ⓕ of ram is machined with such a precision that it is at right angle with sliding portion of ram in both vertical and horizontal planes. This ensures that face Ⓕ of ram is precisely parallel to body circular face where die holder sits. For best extrusion results, it is also necessary that movement of ram in its guideways is also at right angle with body face where die housing rests. These accuracies may be tested by fixing a test bar on ram and suitably fixing a dial indicator. Spindle point of dial indicator should touch test bar diameter with some deflection shown by dial indicator reading. On moving the ram forward and backward, dial indicator needle may or may not deflect. If it does not deflect, it can be concluded that face of ram is at right angle to ram slide. Again, fix dial indicator suitably on the face of ram and a right angle test bar on the face of machine where die housing rests. Both, dial indicator and test bar must be so located that point of dial indicator spindle should travel along the length of test bar. There will be no indication of deflection if movement axis is exactly right angle to machine face where die housing rests.

Extrusion of Can from Zinc Calot

In practice it is found that extrusion of zinc calot is difficult as compared to that of aluminium. This is due to the fact that deformation characteristics of zinc are such that higher impact force per unit area is needed to extrude as compared to aluminium. Zinc calots are generally lubricated by graphite and small amount of suitable oil. This makes extrusion possible with good results.

Economy in Calot Blanking

Many industrial units, producing cans/containers purchase calots from other companies, producing calots of various shapes, sizes and materials. Calots may be of round, octagonal, hexagonal, square and triangular geometrical shapes. Adoption of triangular and square shape is ruled out as the extruded can will be defective. Octagonal shape for calot is also ruled out for two reasons. One, line to line matching in blanking layout is geometrically not possible. With hexagonal geometrical shape it is possible to keep line to line touch in blanking operation. Figure 3.12 shows an aluminium strip from which round calots of, say 38 mm diameter are punched out. Tool is designed to blank out two blanks in first row and three in second row. Since the thickness of aluminium strip is about 6 mm (assumed), a tie of about 3 mm will have to be provided. According to a typical layout, 65 numbers of calots would

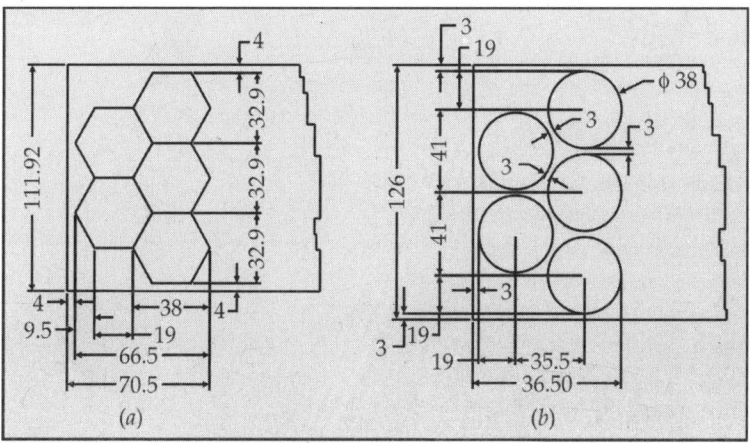

Fig. 3.12: Calot Blanking Layout

be produced in a strip of one metre length. Total area of calots works out to be 737.1 sq cm whereas area of one metre long and 12.6 cm wide strip would be 1260 sq cm. Hence area of scrap would be 522.9 sq cm. This scrap area is 41.5 % of one metre long strip area. Percentage of scrap could have been reduced if shape of blank would have been hexagonal. But making of a tool for blanking hexagonal calots is quite expensive in comparison with blanking tool for round calots. Producer of calot thinks of pros and cons of getting a hexagonal tool made as its maintenance and production life is less as compared to tool life of round calots. But saving due to reduction of scrap would be quite substantial. Hence producer of calots and designer of tool may jointly opt for making tool to produce hexagonal calots.

Since zinc is an expensive material hexagonal calots are generally produced.

Photographs

Photo 3.1

Reverse impact extrusion press is shown from operator's side. Calot feeding system, switching panel, big flywheel having pneumatic clutch and brake system, base thickness adjusting hand wheel, motor and FRL unit can be seen. (*Courtesy: SISL*)

Photo 3.2

Again it is reverse impact extrusion press, shown from opposite of operator's side. This is the side from where extruded components come out. (*Courtesy: SISL*)

Photo 3.3

A view of extrusion line with controls and automation (*Source*: Schuler AG-Germany) .

Photo 3.4

Transfer by dial feed plates, absolute reliability in transport', this is what is shown in the photograph (*Source*: Schuler AG-Germany) .

Photo 3.3

'Conventional impact extrusion presses are equipped with a normal knuckle with three articulated joints. The XS series features four articulated joints in its modified knuckle'. (*Source*: Schuler AG-Germany)

4

Vacuum Metalising of Reflectors

Basic Concept

It is a common observation that washed and rinsed clothes hung on a drying rope get dried up after some time. Time taken by clothes to dry depends on atmospheric temperature, humidity and turbulence in atmospheric air. While drying of clothes is going on, no steam formation is observed, still clothes get dried up. Drying process gets completed fast if weather is dry. This means that humidity is low, say 34 (relative humidity). Drying is still fast if atmospheric temperature is high, say 42 degrees centigrade. Clothes get dried up even in winter when relative humidity is low and atmospheric temperature is also low.

Most of molecules of water get slowly away from each other and material of cloth and get lodged with air molecules. This natural process turns the water into vapours which is invisible moisture in air and it is called evaporation.

Kerosine kept exposed to atmospheric temperature and pressure will turn into invisible gaseous form. If equal volume of water, kerosene, petrol and carbon tetrachloride is simultaneously kept exposed to pressure, temperature and humidity of atmosphere, it may be observed that complete evaporation of various liquids takes different periods of time. Above mentioned examples indicate that rate of vaporization is different for different liquids in same environmental conditions. Another glaring example is of liquefied petroleum gas (LPG). It is kept liquefied at about nine times the atmospheric pressure. If the pressure of confined gas is allowed to drop, LPG will immediately start to vaporize and increase the pressure in container (gas cylinder) till the vapour pressure of LPG is balanced by pressure inside cylinder. Vapour pressure of any liquid is the pressure exerted by the molecules to attain state of vapour.

Evaporation of Metals

Metals and their alloys do not evaporate at normal atmospheric temperature and pressure. They start evaporating if their temperature is risen to such a value that its vapour pressure increases beyond chamber pressure, which is usually maintained at 10 to power minus 4 to 10 to power minus 5 torr. When an evaporant is heated in vacuum, its thermal energy is increased to such a value, say 0.08 to 0.35 eV, that molecules (atoms) are forced out from the surface of evaporant. Forced out atoms get dislodged and move away from metal evaporant in a straight path. These travelling atoms come across with gas molecules which are left out from evaporation of gases, moisture, etc. from components in the vacuum chamber and chamber itself. When atoms of evaporated metal collide with left out gas molecules, movement of metal atoms becomes random. So atoms of evaporant in the form of vapour move around and fill the chamber in the form of 'clouds'.

Vapours, when come in contact with cold surface of substrate (component) and inside of vacuum chamber surface, get deposited in the form of very thin layer of metal which is being evaporated.

Expression of Thickness

Ultra thin thickness like that of metalised surface is expressed by a unit called Angstrom. Its symbol is Å. One microinch is equal to 250 Å. It is such a small unit of length (measurement of a thickness is also a length) that cannot be measured by conventional measuring instruments. Sophisticated instruments are used which are normally not used on shopfloor. In practice, thickness of metalisation is assessed by visual check. A glass slide is kept in the chamber at a pre-decided location. After completion of vacuum metalising cycle, slide is taken out and viewed against light. If the light is faintly visible then it would mean that thickness of deposit is satisfactory. If more light is visible then the conclusion may be drawn that thickness of deposit is much less. It is a visual comparison and primarily depends on the experience and judgement of examiner.

An inexpensive electronic comparator can be built where the amount of light passing through the metalised glass slide may be seen on an analogue or digital indicator. It may be calibrated to show the light passing as percentage of light passes through uncoated (unmetalised) glass slide.

Description of Vacuum Metalising Plant

Vacuum metalising plant suitable for metalising sheet metal and plastics reflectors by aluminium evaporant consists of following equipments.

- Vacuum chamber
- Evaporant heating system
- Reflectors holding and rotating system
- Rotary vacuum pumps
- Oil diffusion pump
- High pressure air
- Various valves
- Water chilling plant
- Water circulation pump
- Auxiliary solenoid valves
- Wiring and piping
- Control panel
- Reflector carriers

Apart from this, room or hall should be according to statutory laws governing ventilation, movement space, entry and exit provision, roof height, etc.

Plant Room

Vacuum metalising plants of various capacities have equipment of different dimensions. These equipments together with main equipment, that is vacuum chamber, are to be installed in room or small hall. Length, width and roof height should be such that complete plant gets accommodated and there remains enough space for movement, component carrier stand movement, trays, some tables, chairs, almirah, etc. Service equipments like water chilling unit, water pump and air compressor may be installed in an adjoining shed. Plumbing from chilling plant and air compressor may be laid through the wall of room to reach main plant. Piping from outlet of vacuum pumps can be laid to pass through wall of room to open air. This is to discharge gases, fumes, etc. to open air so that inside of room is completely free from exhaust gases. An important factor to obtain best results of vacuum metalising of reflectors is clean and dust-free environment in the room. Air in the room should be free from suspended particles of dust. This can be achieved by arranging injecting fresh filtered air of such a volume and pressure that pressure of air inside air-conditioned room is slightly higher than outside atmospheric pressure. This will ensure to a great extent that dust particles suspended in atmospheric air do not enter the room. Persons coming inside room should remove their shoes outside the double door entry/exit way. There should be a conveyor through a wall opening to bring reflectors inside room from adjoining moulding shop or baking room. Temperature inside vacuum metalising plant room may be maintained at a comfortable level so that operators or helpers working do not sweat.

Figure 4.1 shows a typical layout of plant. Equipment, machinery and attachments shown in figure are listed.

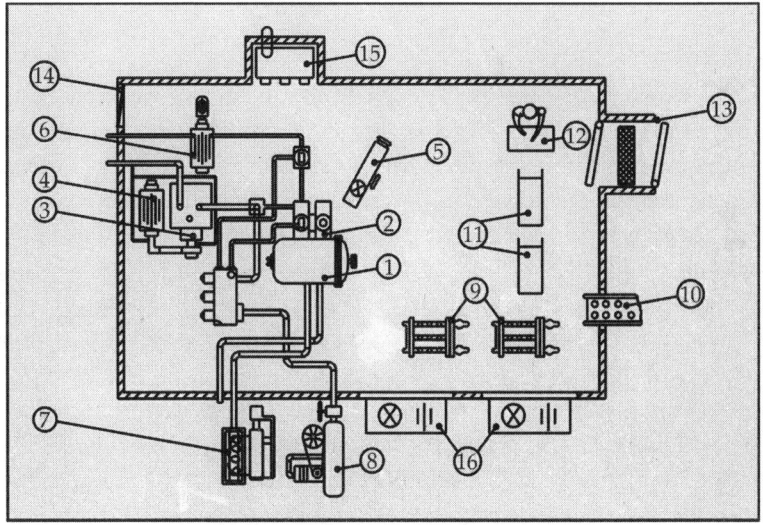

Fig. 4.1: Typical Plant Layout of Reflectors

1. Vacuum chamber
2. Oil diffusion pump
3. Fore pump
4. Back up pumps
5. Control panel
6. Passage valves
7. Water chilling plant
8. Air compressor
9. Reflector carriers
10. Conveyor belt
11. Reflector packing station
12. Attendant seat
13. Double door entry/exit
14. Emergency door
15. Electrical main control panel
16. Air conditioner

Vacuum Chamber

Vacuum chambers are generally in the shape of bell jar. Depth and bore of chamber varies according to designed capacity of machine. For small capacities of vacuum metalising plant, bore of chamber may be around 60 cm and depth 70 cm. Bigger capacity plants may have chamber of 100 cm bore and 125 cm deep. These dimensions may be more. Various manufacturers of vacuum metalising plants design and make chambers of various combinations of bore and depth dimensions. Chambers are manufactured from ferrous or aluminium alloys which resist formation of rust which generally takes place with iron and steel exposed to atmospheric moisture and heat over a period of time. Further, wall thickness of chamber (bell jar), together with its flang is kept so much that there is no chance of its collapsing when high vacuum is generated inside the chamber.

In horizontally designed vacuum metalising plants there are generally two ways of placing bell jars in position and removing them. Basic concept of two systems is illustrated by Fig. 4.2 (*a*) and (*b*). Some manufacturers for smaller plants adopt bell jar removal and replacement arrangement as shown in Fig. 4.2 (*a*). There is an angle iron structure (stand) on which suitable size of guide rails are fitted. Bell jar has small wheels. Distance between coaxial wheels is kept so much that suits guide rails. Bell jar can be pulled or pushed easily. Upper structure with railing is such that it can be swung more than 90 degrees at one side with bell jar on it. Now rotating jig, fitted to machine, is manually loaded with reflector carriers. Now the trolley with component carrier placed on it can be brought in position to push carrier so that it gets locked with holding device and rotating driver gear. After the carrier is placed in position, bell jar is swinged back in position and pushed over the railing till rubber O ring, fitted to bell jar flang face, touches the plain surface of fixed housing carrying gear revolving system and electrodes for placement and heating of filaments. A swinging handle is pulled to lock the bell jar with housing with a little pressing on O ring. This arrangement does not allow air to sneak in when evacuation of air inside bell jar is taking place.

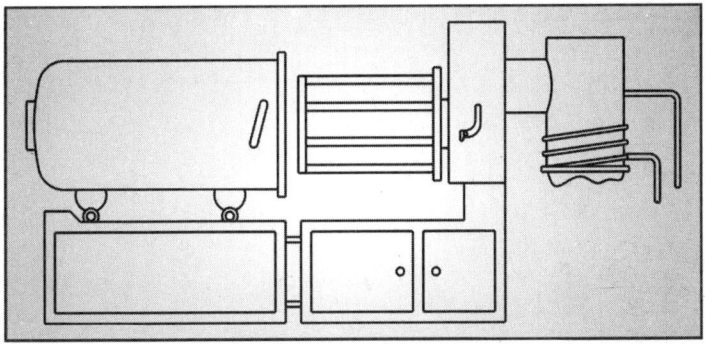

Fig. 4.2 (a): Swinging Chamber

Figure 4.2 (*b*) shows an arrangement where chamber is in two parts. Cylindrical part permanently fixed to machine block and bulged shape door swings open or close over the cylindrical part of chamber. Door carries a reduction gear motor to rotate carrier when loaded inside. Door is also fitted with inspection glass window. Inner side of inspection window has an arrangement to put extra glass disc with ease and to remove it after metalising process is over. Air is allowed to enter the chamber through an air vent which allows controlled flow of air and not all of a sudden.

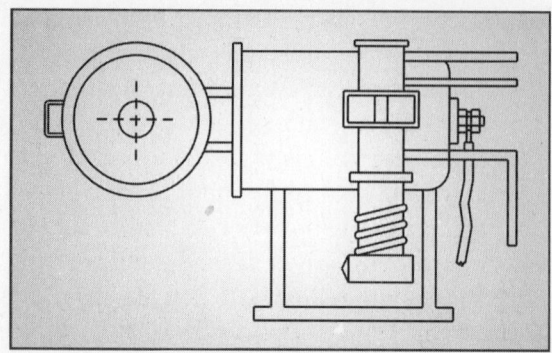

Fig. 4.2 (b): Swinging Door

Figure 4.3 is a typical sectional view when seen from top. Enlarged view of fitting of copper bus bar to chamber wall is also shown.

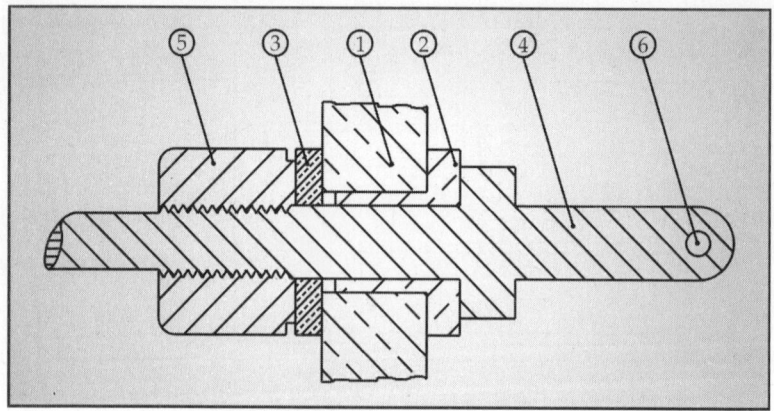

Fig. 4.3: Bus Bar Fitting

Referring to Fig. 4.3, ① is chamber wall which has two through holes at suitable PCD. ② is a teflon bush very precisely made and having smooth surface. ③ is a good quality teflon washer. ④ is copper bus bar. Actually there are two. ⑤ is hexagonal nut of fine threads. ⑥ are clamp arrangement for clamping electric cable.

Evaporant Heating System

Filaments are three core twisted tungsten wire on which evaporant aluminium wire pieces are hung (Fig. 4.4).

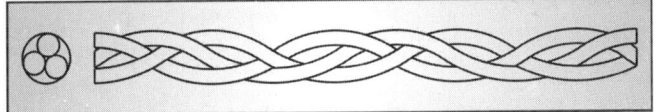

Fig. 4.4: Filament

Tungsten is used to make filament because it will not evaporate at temperature which is required to vaporize aluminium in a vacuum of 10^{-4} or 10^{-5} torr. Tungsten would vaporize at about 3200 degrees centigrade in a vacuum of around 10^{-2} torr. Aluminium would vaporize at about 1200 degrees centigrade in a vacuum of 10^{-2} torr.

Whole assembly, as shown in Fig. 4.3, is done with great accuracy and cleanliness to ensure leakproofness. Copper bus bars have holes ⑥ to tighten electric supply cables from step down transformer. Bus bars are usually supplied with voltage which can be varied, say from 5 to 25 volts. Current drawn by heating filaments may be as high as 100 to 200 amperes, depending upon the number of filaments clamped and the voltage applied. By increasing or decreasing the voltage applied to primary of transformer, output voltage of secondary terminals may be varied.

Increase or decrease in the temperature of filament may be done manually by varying primary voltage of transformer by mèans of a rheostat. Change in voltage may be read on a voltmeter fitted on control panel. There are sophisticated vacuum metalising plants. Control of various process parameters is automatically done with inbuilt sensors and microprocessor in control panel.

Oil Diffusion Pump

Oil diffusion pumps are capable of evacuating metalising chamber to a vacuum of 10^{-3} to 10^{-5} Torr and even more. Working principle of oil diffusion pump is explained with the help of Fig. 4.5. ① is cylindrical body of oil diffusion pump. ② is oil reservoir having a capacity of about two litres of oil for a small plant. ③ are electrical heaters, fitted under oil reservoir to heat oil to vaporisation temperature. ④ are baffles to catch condensing molecules of oil when oil vapours get cool on rising up. While the vapours are up, molecules of air and gases get attached to oil molecules. Condensed vapour of oil brings down molecules of air and gases and touches baffle surface. Oil vapours on getting in contact with baffle surface condense back to liquid oil. In this process of turning from vapour to liquid, molecules of air and gases get detached and sucked by back up mechanical pump. In this way evacuation of air and gas molecules keeps on taking place. Consequently, a high vacuum of 10^{-5} torr can be achieved in vacuum chamber. ⑤ are coils of hollow pipe tightly wrapped around cylindrical body. Chilled water is circulated through pipe coil. Temperature of chilled water at entry point is maintained around 10° to 5° C. As a result of circulation of chilled water, walls of cylindrical body remain cool which is necessary for oil vapours to condense.

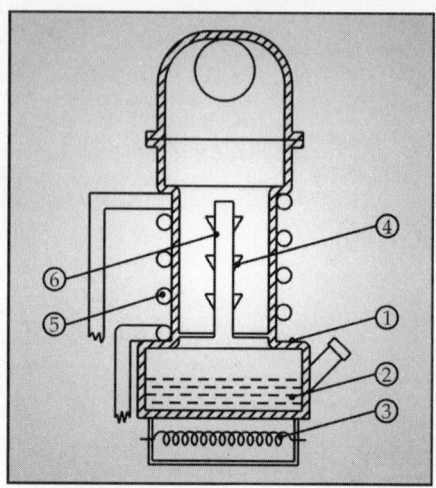

Fig. 4.5: Oil Diffusion Pump

Quality of oil should be such that it does not get disintegrated on being continuously heated. Oil should be very pure. Molecules of oil in vapour state should attract molecules of air and gases. Specially formulated silicon based oils are used. These oils are quite expensive and should be used with care. ⑥ are baffle projections which are supported by a central column. Projections are tilted with a suitable angle so that condensed oil trickles down in oil reservoir.

Lower portion of cylindrical body has a pipe of suitable size nicely welded, that is, without any leak. This valve is connected to back up rotary vacuum pump which keeps on evacuating air and gas molecules which are pumped in from vacuum chamber. Time taken by oil diffusion pump to create a vacuum of 10^{-5} torr in metalising chamber depends on following factors:

- Size of chamber
- Cleanliness and vacuum tightness of various joints
- Purity of evaporant, say aluminium
- Material of substrate
- Capacity of fore pump, online diffusion pump and back up pump

Size of vacuum metalising plant chamber basically depends on the dimensions and quantity of components which are required to be metalised in one charge, depending on requirement of production quantity per eight hour.

Processor selects a vacuum metalising plant which has a chamber size to suit production output requirement. Makers of vacuum metalising plants have standardized few set of dimensions of chamber. Customer (prospective purchaser) has to select a vacuum metalising plant, chamber of which suits the customer. Makers of plant normally do not make a chamber other than standard dimension. Various standard inside dimensions and depth have a range of 70 cm diameter and 100 cm depth to 200 cm diameter and 250 depth and even more.

Design of complete system of vacuum metalising plant depends on chamber volume, intended evaporant, time limit for completion of metalising of one charge.

Therefore, capacities of fore pump, back up pump are determined accordingly. Fore pump is generally single stage vane type pump. It is capable of attaining a vacuum of 10^{-1} or better. Capacity of pump is selected to suit initial evacuation of air and gases from the chamber of a perticular volume with a specific time. Basic construction of a vane type single stage pump is shown in Fig. 4.6. Referring to Fig. 4.6, ① is a closed grain cast iron

Pump body block: It is casted with such technique that there is no chance of air sneaking through possible pores in the body while vacuum is built up inside. ② is rotor having precision machined slot. Normally slots are in odd numbers, generally 5 or 7. ③ are vanes of brass, teflon or thermoset plastics impregnated sheet. On some occasions CI vane are also used. Most important point is that vanes should slide in slots with precision. A running fit gives satisfactory results. Since the rotor and vane are operating in developed vacuum, there is a chance of leakage of air or gas through clearance between vane and slot but the presence of oil film prevents leakage. Rotor diameter and faces are precision machined. Bore of pump body is also free from ovality, taper, surface undulation, roughness, etc. Rotor is held inside bore of pump body by means of side plates. Rotor shaft passes through one side plate. Stator (bore of pump body) and side plates are so precisely and rigidly fitted that there is a line clearance of 0.01 to 0.025 mm between rotor and stator. Lubricating oil present in the pump keeps clearance sealed against sneaking of air from exhaust side of seal to sucking side of seal. Similarly gap between the rotor faces and side plates are also precisely maintained. This ensures good efficiency of pump achievement of good vacuum. ⑤ is exhaust port in pump block ① and ⑥ is inlet (sucking) port. ⑦ is a housing for a plate type non-return valve ⑧. ⑨ is a pipe attached to exhaust port. In turn this pipe is connected to other pipes which opens outside the plant room.

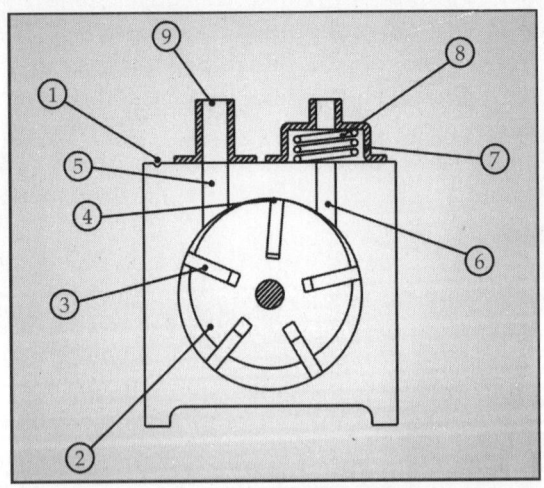

Fig. 4.6: Fore Pumps

Extreme cleanliness during assembly of pump, care and maintenance during service ensure good performance of pump.

Back up pump is a two stage pump which is capable of producing vacuum up to 10^{-2} torr or better. Design and construction of back up pump are almost the same as of single stage pump. Only difference is that in two stage pumps, two pumps of single stage are joined together in series.

Rotary vane type vacuum pump if gets rotated in wrong direction, may get damaged and also cause harm to vacuum system. Wrong rotation of vacuum pump is prevented by fitting unidirectional roller bearing between rotor shaft and face plate.

Purity of Evaporant Material

Light reflectors are generally vacuum metalised by aluminium. Quality of reflector and brilliance of metalised surface depend on a number of factors, including purity of aluminium evaporant, which is a wire in shape of V. 99 to 99.7% pure aluminium wire practically gives satisfactory results.

Presence of minute quantities of other elements like bismuth, cadmium, antimony and other compounds may be termed as 'impurities'. Presence of these impurities beyond 0.3% interferes with the quality of metalising. Reason is that partial pressure of impurities is more than chamber pressure at temperature below required to vaporise aluminium. Cleanliness of aluminium wire evaporant bits is also necessary to avoid evaporation of moisture, sweat, traces of grease, oils, etc. Vapours of impurities interfere with the deposition of aluminium vapour on surface of substrate, poor metalising quality is achieved.

Requisites for Good Reflectivity and Brilliance of Reflectors

One of the requisites is that surface to be metalised should be facing towards aluminium bits ① which are to be evaporated. Fig. 4.7 (*a*) shows an arrangement of reflectors ③ on a carrier ② which is a rotating jig around evaporant. For small capacity plants (say, 100 cm chamber diameter) distance of surface to be metalised from source of evaporant should be around 20 cm (distance r_1). Jig of Fig. 4.7 (*a*) may be rotating or non-rotating type. Reflectors are loaded on carriers in two circles. Inner circle where surface to be metalised is facing aluminium bit from where aluminium would vaporise when heated. Atoms of aluminium in vapour would directly travel radially outwards to form 'clouds' of vapour in the chamber. Hence the facing surface would get deposited by aluminium. Set of reflectors held in outer circumference, facing towards chamber walls would not receive vapours of aluminium properly. Therefore deposits of aluminium would be uneven, insufficient and not at all. Therefore there is no use of fixing reflectors in outer circle but this is an unacceptable situation from productivity point of view. One charge of metalising would yield less components. To overcome the disadvantage of less output, carriers with rotating jigs are designed and made to have 'planetary' movement as well. In this way all loaded reflectors keep on rotating to face source of vaporisation during vaporisation time of few minutes. This arrangement is shown in Fig. 4.7 (b).

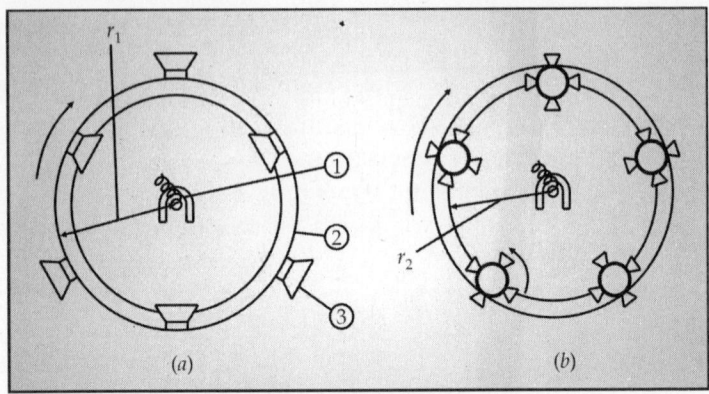

Fig. 4.7: Reflector Carrier

Passage Valves

Fore pump, back up pump and oil diffusion pump are connected to vacuum chamber through a number of valves. These valves are meant to open or close the passage for evacuation of air or gases in vacuum chamber. Construction of valve is explained with the help of Fig. 4.8.

① is a circular passage which goes to oil diffusion pump. ② is a round plate having a circular groove for placement of a rubber ring ③. ④ is ball joint which keep round plate floating a little. Object of making a ball joint is that plate is allowed to sit properly on face ⑮ of passage ①. Ball rod passes through a vacuum tight seal ⑤. ⑭ is the opening of valve housing in vacuum chamber. ⑥ is a flexible joint to join ball rod to rod ⑦ of pneumatic jack ⑧. ⑩ is a pneumatic directional control valve which allows high pressure air to enter in pneumatic cylinder through pipe ⑬ and to provide passage to returning air from top of the piston of pneumatic jack when solenoid valve ⑪ is energized, passage in directional control valve changes, allowing high pressure air to enter in space in pneumatic jack above its piston. In this way round plate ② moves down to close and open passage between oil diffusion pump and vacuum chamber. Similar type of valves are fitted between chamber and fore pump and between oil diffusion pump and back up rotary pump.

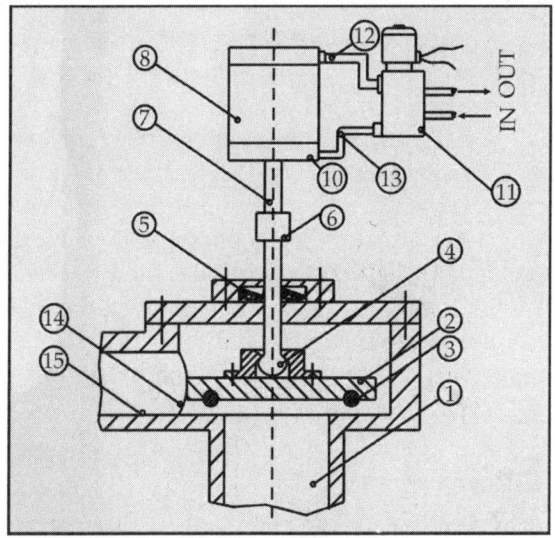

Fig. 4.8: Construction of Valve

Water Chilling Plant

Water chilling unit is installed outside vacuum metalising plant room. Chilled water is continuously supplied to oil diffusion pump cooling coil. Chilling plant consists of refrigeration system which constantly cools water to about 10 degrees centigrade. Cool water is pumped through piping, fittings and cooling coils of oil diffusion pump back to chilling plant. Capacity selection of water chilling plant depends on the capacity of vacuum metalising plant. A small plant may need about 10 litres of chilled water per minute flowing through cooling coil of oil diffusion pump.

Air Compressor

Generally two stage compressor is used for actuating pneumatic jacks of vacuum metalising plant. Two stage compressors are more efficient as compared to single stage compressor. Reciprocating piston type compressor is mounted on a suitable strong tank in which compressed air is stored. Reciprocating piston type compressors are capable of producing pressure in reservoir as high as 100 kg/cm^2 with capacity of about 600 litres per hour at 70 kg/cm^2. Compressors are fitted with pressure release sealed valve which does not allow pressure to exceed a fixed value. In addition to this, a pressure safety valve is also provided to release excessive pressure of air in case pressure release valve fails to operate. There is also an electromagnetic switch for loading and unloading compressor automatically.

There is a provision to drain out water which gets condensed in tank and pipe line. Drain cock is provided near the joint of pipe to tank. It is very necessary that air free from moisture should go in the pipes. FLR (filter, lubricator, regulator) units are fitted between high pressure air supply and directional control valves. Use of FLR unit allows air, filtered and laden with lubricant molecules, to pass with a preset pressure to directional control valve. In operating manual of plant, working air pressure is generally specified which is to be set. In practice it is around 10 kg/cm^2.

Reflector Carrier

Reflectors are first fixed on carriers which are attached to a specially designed jig. Figure 4.9 shows basic construction of jig and carrier.

(1) is a trolley type construction made of non-rusting steel. This trolley has two vertical frames (2) and (4). (2) is in the shape of ring which carries a number of spring loaded carrier holding devices (11). These devices are pulled by hand to create gap for putting carriers (12) in place and then devices are released to hold carriers without any chance of getting out of place during its functioning in vacuum chamber. It may be noted that ring (2) is made to have big bore. This is provided for bus bar (fitted in centre inside the chamber) to get in. Other vertical frame/plate (4) consists of carrier holding spindles with sprockets (6) attached. There is also a sprocket (3) which is attached to trolley type construction (1). All the sprockets are in one plane. A chain (5) passes over all the

gears as shown. A spindle fixed to frame/plate ④ comes out through bush bearing of trolley's vertical portion. At the end of spindle is fitted a circular disc ⑨ which has a slot. If disc ⑨ is rotated by hand, both vertical plate ④ and vertical ring ② will also rotate. Since sprocket ③ is stationery, chain will have a motion when plate ④ gets rotated. In turn, chain would make gears ⑥ to rotate. Carrier holders would also rotate. In this way all carriers rotate in a planetary motion. It is worth noting that vertical ring ② is supported on two rollers ⑬ for rotary motion with vertical plate ④ as both are rigidly connected with bars ⑭.

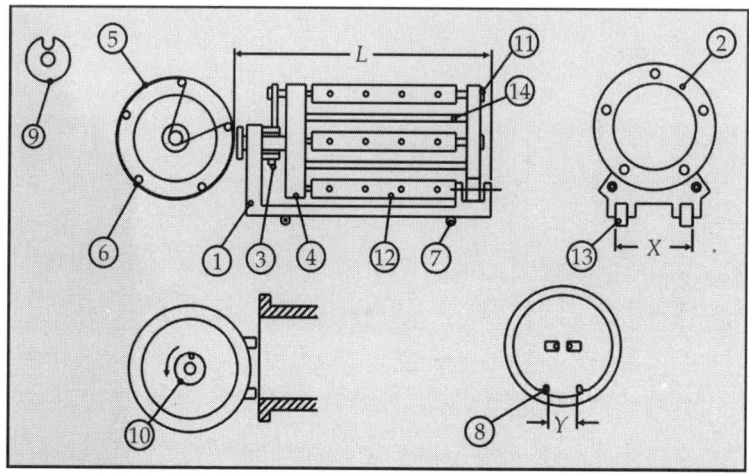

Fig. 4.9: Jig and Carrier

There are four travelling rollers fitted under the jig. Inner distance (X) between the rollers is more than (Y), the distance between the rails fitted (welded) in the chamber. Overall length (L) and effective radial dimensions are so kept that jig enters vacuum chamber easily. A motor with reduction gear is fitted outside the chamber door to rotate disc ⑩ which has a small protruding pinion. Before closing the door slot of part ⑨ is positioned so that on closing the door, pinion enters the slot. Now on starting the motor, carrier and jig inside vacuum chamber start rotating. As explained earlier, carrier rotate in planetary pattern. Carriers are so designed and made that substrates are firmly held so that they do not get dislodged during rotation.

Shape and dimension of holding devices to hold reflectors on carriers depend on shape and dimensions of reflectors. Reflectors have a variety of hole or neck shape and dimensions. Figure 4.10 show few typical reflector holes and necks.

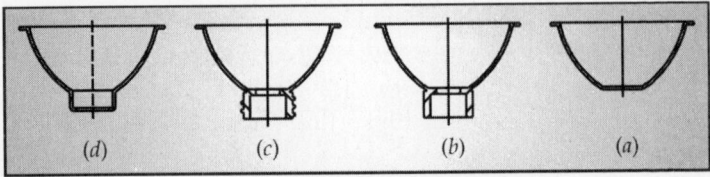

Fig. 4.10: Typical Reflector

Reflector shown in Fig. 4.10 (*a*) can be held in three ways:

- By three-legged clip made from spring hard brass sheet to hold reflector from hole.

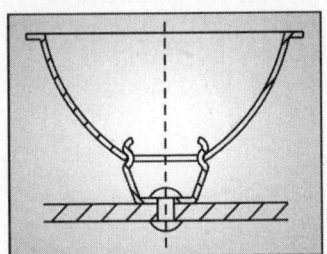

Fig. 4.10 (a): Reflector Holding

This type of clip may not be suitable as there is a chance of 'shadow' (dark spot) of clip legs on reflecting surface which is a defect.

- Other possibility is to hold the reflector on the collar as shown in Fig. 4.10 (*a*) 2.

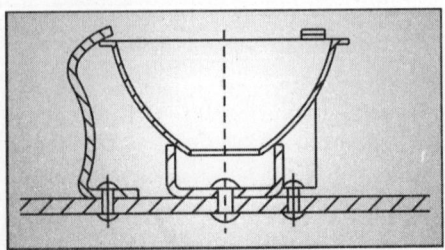

Fig. 4.10 (b): Another Reflector Holding

In this arrangement 'shadow' from three clips may form on collar which is normally kept hidden in assembly of a lighting device.

- Sometimes, any deposition of aluminium film on back side of reflector is undesirable due to interference in electrical circuitry or aesthetic reasons. Under such circumstances masking of back side of reflector can be done. Such a possibility is shown in Fig. 4.10 (*a*) 3.

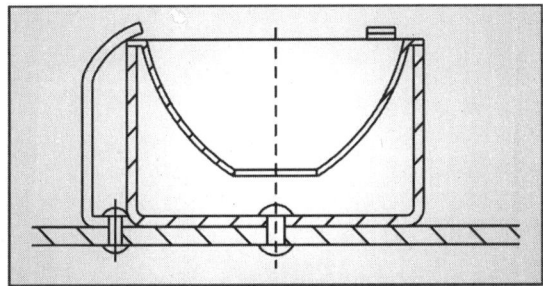

Fig. 4.10 (c): Masking of Reflector

Number of Reflectors in One Charge

Number of reflectors which can be loaded in one charge depends on collar diameter, height, number of carriers on the jig, length of carrier.

Let there be three types of reflectors having following dimensions

Table 4.1		
Type	*Collar diameter* ϕ	*Height* h
S	20	15
M	38	30
L	100	45

Carrier is hexagonal in shape. Centre of carrier from inner wall (*W*) of chamber is (*S*). This is a fixed distance for various loading between jigs and inner diameter of chamber. Let it be 90 mm.

Length of carrier = 150 cm

Number of carriers = 5

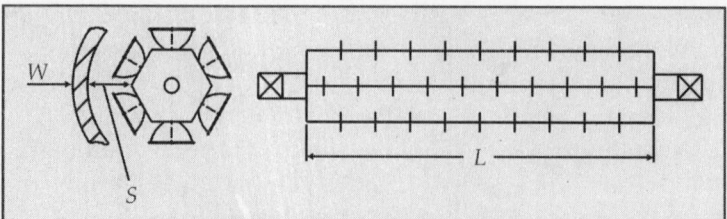

Fig. 4.11: Hexagonal Carrier

Type	Nos. in one row of carrier	Nos. in six rows of carrier	Nos. in jig
	Table 4.2		
S	50	300	1500
M	31	186	930
L	12	36 (three rows)	180

Production of reflector may be around 360 numbers if it takes 20 to 30 minutes to complete one charge. Completion time means time from pushing the jig trolley into the chamber to taking it out after metalising and allowing the air to enter chamber to open the door.

In case production requirement for large reflector is more then a bigger capacity metalising plant may be needed.

Functional Block Diagram

Referring to Fig. 4.12, C is vacuum metalising chamber. P_1 is roughing pump. V_1, V_2 and V_3 are pneumatic valves. Vd_1, Vd_2 and Vd_3 are directional control valves. These valves control entry of high pressure air to pneumatic valves and exit of return air from V_1, V_2 and V_3 respectively. All the three valves V_1, V_2 and V_3 are normally closed. This means the passage for suction to vacuum pump P_1, P_2 and P_3 is closed. These valves open the passage only when pneumatic jacks of valves get actuated by entry of high pressure air.

Two pressure sensors are fixed in vacuum line system. One pressure transducer (sensor) is S_1 which is fitted between oil diffusion pump and valve V_3. This transducer is connected to high vacuum indicator on control panel. It starts sensing when vacuum of 10^{-3} Torr is attained and can indicate up to 10^{-7} torr or better.

Other pressure transducer is S_2 which is fitted between chamber C and valve V_1 and V_3. It is connected to another vacuum indicator. This indicator indicates vacuum from 10^{-1} to 10^{-3} torr.

V_4 is an air vent to chamber which is opened manually to allow air to enter chamber. In automatic plants it is controlled automatically.

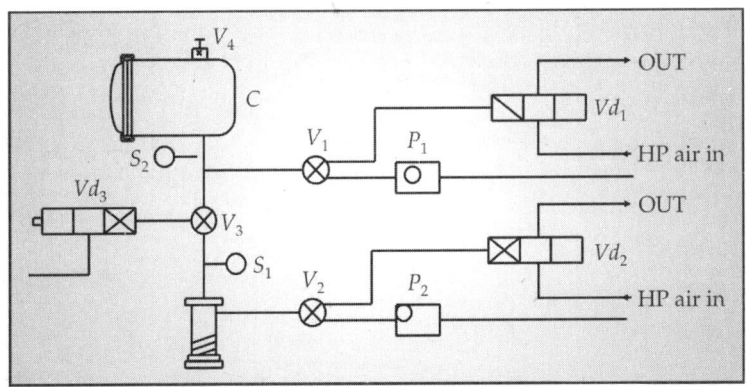

Fig. 4.12: Functional Block Diagram

Control Panel

Purpose of control panel is to carry all switches, indicators, rheostat, audiovisual warning lights to control operation of vacuum metalising plant. Control panel of an automatic plant may have microprocessor and digital display of various parameters. For a particular product, values of various process parameters may be entered and saved in the memory of computer fixed to control panel. In spite of automation, operator may take over the control partly or fully as and when required. Example of such situation is heating of evaporant. While heating of filament is in progress, operator observes from observation window to see if heating of aluminium wire bit is going on properly. If operator, by visual judgement, feels that temperature of filament is less then he manually increases current passing through filaments to increase its heat. Reduction in heat may also be sometimes necessary.

Automatic plants may have facility of process parameter data collection while production is going on charge after charge. It may be possible to take printout of collected process parameter data.

Plant Operation Sequence

Operation of vacuum metalising plant with proper sequence is almost a must for achieving good results of metalising. Before starting the plant following may be done on daily basis.

- Cleaning of room/hall, removal of dust, fallen bits, etc.
- Cleaning of all carriers, jigs and trolleys
- Removal of unwanted or rejected components from room
- Making furniture, tool box, etc. tidy
- Checking of oil levels in all vacuum pumps
- Drain out condensed moisture from high pressure air line
- Check FLR (filter, lubricator, regulator) unit for presence of lubricating oil to desired level and drain out water from cup
- Switch on chilling plant and air compressor
- Switch on fan of filtered air to enter into room
- Switch on air conditioner (optional) if necessary
- Wait for some time till switched on machinery gets stabilized
- In the meantime start loading carriers with reflectors

Now the sequence of operation of plant may be as follows for a manually operated plant (this is a typical example. Sequence may vary from plant to plant. It is advisable to follow instructions which are normally given in Instruction Manuals of particular plant).

- Switch on roughing fore pump
- Open valve of high pressure line
- Open valve of chilled water to start circulating through cooling pipe coil of diffusion pump
- Fix new heating filaments on bus bars in the chamber
- Hang U-shaped aluminium wire bits. Number of bits to be hung should be predetermined
- Push the loaded jig inside the chamber
- Close the chamber door with the care that pin of rotating driver system gets engaged into the slot of driven jig plate
- Activate roughing line valve V_1 by pushing 'push to on' button on control panel
- Activate fore pump line valve V_2 by pushing 'push to on' button on control panel

- When low vacuum indicator on control panel shows vacuum better than 10^{-1} torr then switch on heating of oil diffusion pump
- On one side evacuation of gases, air from chamber is taking place and on the other hand air in oil diffusion pump is being evacuated by fore pump. After few minutes of putting the heat on, diffusion pump starts to evacuate and putting down air molecules present in oil diffusion pump
- When a vacuum in oil diffusion pump is indicated as 10^{-2} torr, activate valve V_3 to open thus connecting oil diffusion pump to chamber. Electrical circuitry is such that on opening V_3, V_1 automatically closes, thus isolating roughing pump from chamber
- Wait for few minutes for vacuum in the chamber to rise to better than 10^{-4} torr
- Start rotation of jig by pushing 'push to on' button on control panel
- Switch on electrical supply to heating filaments
- Start increasing the flow of current through filaments gradually while seeing inside the chamber through observation window. For few moments visible heat of filaments can be seen
- Keep on gradually increasing the heat in filaments till aluminium starts evaporating and inside glass of observation window gets metalised and seeing inside gradually becomes difficult, depending on thickness of deposit of aluminium
- Let the process of evaporation continue for predetermined time of few seconds (even a minute or so)
- Rotate back rheostat to bring down supply of current and switch off electrical supply
- Stop rotation of jig
- Close valves V_3 and V_1
- Admit air by manually operating air vent valve
- In a short time air pressure in chamber would be the same as that of atmosphere
- Open the door
- Pull out jig

- Check quality of metalising in reflectors of all carriers
- If quality is correct then reflectors should be unloaded from carriers and placed in cardboard boxes by person other than operator
- Repeat process by using other jig which is ready with loaded reflectors for placing inside the chamber

Quality Aspects

Quality of vacuum metalised reflectors depends on following aspects:

- Quality of reflecting surface of substrate, that is reflectors
- Material of substrate
- Quality of evaporant (aluminium)
- Accuracy of metalising process

Reflectors are generally produced either of brass and aluminium sheet metal or plastics. Reflectors produced from sheet metal undergo a number of surface preparations to make it suitable for vacuum metalising. Depending upon grains opening of reflecting surface, mechanical buffing is done to achieve smooth and polished surface. Since polishing is carried out with polishing material, some microparticles of polishing compound remain impregnated in micro grain openings of sheet metal. These microparticles of polishing compound are removed by chemical washing and bright dipping. In spite of all above mentioned treatment, surface of reflector does not become ideally smooth and lustrous. So, to achieve lustrous surface, it is covered with specially formulated lacquer. Reflecting surface is generally lacquered by spraying. Lacquered reflectors are then placed in baking ovens for a specific time and temperature. After baking is completed, reflectors are taken out and allowed to cool down to room temperature. Now lacquered surface is well adhered to base surface. It is very smooth and shiny. Now reflectors are ready for vacuum metalising.

For plastics reflector no surface preparation is generally essential. Plastics reflectors are generally moulded by injection moulding process. Moulds consist of cores and cavities made of special steels. Parabolic portion of cores are polished to extremely high smoothness and shine. Moulded reflectors have smooth and shiny reflecting surface. Formation of focus spot of light can be checked even before metalising. This is to ascertain if parabolic profile is accurate. Once these reflectors are vacuum metalised, a highly smooth and brilliant surface is achieved.

Selection of plastics raw material for moulding reflectors is very important. Type and grade of plastics should be such that chemicals like plasticizers do not migrate out of plastics 'body' to surface when residing in high vacuum environment, that is metalising chamber of vacuum metalising plant. Many types and grade of plastics may be found suitable. But from cost-effectiveness point of view a compromise may be made in the selection of plastics. General purpose polystyrene gives very good shiny surface but it is brittle. If component is likely to be subjected to impact during its service then general purpose polystyrene may not be a correct choice of material. General purpose polystyrene may be selected if the component is protected by other components in the assembly. Medium impact and high impact polystyrene are impact absorbing, unlikely to break as compared to general purpose polystyrene. But due to presence of impact improving polymers or rubber, its use for vacuum metalising is limited. Due to evaporation of plasticizer, etc. cycle time for obtaining a vacuum of 10^{-5} torr may take longer time, thus affecting productivity of vacuum metalising plant. Further, reflecting surface obtained is not as shiny as of general purpose polystyrene. Acrylic, ABS or polycarbonate may also be used which are expensive plastics as compared to general purpose polystyrene but have much superior mechanical and surface luster qualities.

Heating filaments are made from a number of wires twisted. This twisted shape of filament helps in wetting (flow of aluminium on it) when starts melting. This may be called wetting of filament. Filament strands are of tungsten which has partial vapour pressure more than 10^{-2} torr at 3200 degrees centigrade. Aluminium vapourises at 1100 degrees centigrade in a vacuum of 10^{-2} torr. Any impurities in tungsten wires may evaporate first and get deposited on substrate as contaminant. Similar is the case with aluminium wire. Impure aluminium wire does not give good metalised surface.

Accuracy of metalising process means proper operation sequence, attainment of correct value of vacuum in specified time, no presence of impurities in system, reliability of vacuum seals, reliability in operation of pumps and passage valves.

Table 4.3: Defects, Causes and Remedies

S.No.	Defects	Causes	Remedies
1	No coating	No heating of filaments	Remove malfunctioning of electrical circuit
		Evaporant (aluminium bits) dropped down	Heat filament slowly
		No vacuum	Correct the system
2	Thin coating, light can be seen across reflector	Less quantity of evaporant (aluminium bits)	Increase aluminium wire diameter and/or number of bits hung on filament
3	Partial coating	Planetary motion of carrier is not taking place	Put on the driving motor Remove mechanical malfunctioning
4	Blackishness	Presence of contaminants, especially oil or grease. Overheating of filaments	Carry out detailed proper cleaning of chamber fitting, filaments and aluminium bits. Control heating current.
5	Adhesion of coating on reflector poor	Reflector surface is having microparticles of moisture and dust, etc. Degaussing during metalising	After moulding remove electrostatic charge from reflectors Choose correct type of plastic
6	Poor shine and brilliance	Base surface lack lustre/ smoothness and shine Use good quality lacquer	Improve surface polish of mould cores to mirror brightness Lacquering and transport environment should be free from dust particles
7	Metalising at unwanted place	Masking of surface is not there	Use properly designed mask
8	Finger marks on reflecting surface	Touched by thumb and fingers	Take precautions that reflecting surface is not touched

5

Miniature Lamp Making Technology and Process

Introduction

A number of varieties of tiny and small incandescent lamps come into the category of miniature lamps. Lamps for dashboards, indicators, flashlights, decoration, etc. are a few examples. In principle, technology of making these lamps is almost the same but there may be a lot of difference between shapes, design, dimensions, machinery and equipment.

This chapter covers some details of technology and process for production of miniature lamps as used for portable light source. There are a number of types of miniature lamps to suit different types of design of light source. Major types are pre-focused, focusing and spot focus (photo 5.1). For these three types of miniature lamps basic technology is almost the same but the shape and sizes may be different.

Bureau of Indian Standards has standardized optical, electrical, mechanical and dimensional specifications. This has brought in quite a uniformity in miniature lamps produced by a number of tiny, small, medium and large scale industries throughout India.

BIS for Miniature Lamps

It would be advantageous for readers of this book to go through BIS. Following parameters of lamp design are standardized.

- Voltage rating
- Current rating at specified voltage
- Light output in lumens per watt
- Light output after half the lamp life
- Life of lamp in hours
- Dimensions of glass bulb and metallic cap

Rated voltage is the voltage which should be applied to a particular lamp. For a range of standardized voltage BIS can be read. Generally, 1.1, 2.2, 2.5, 3.8, 4.2, and 6.2 volts lamps are produced. Following table provides a general information on rated voltage and number of cells/batteries (dry cells of zinc-carbon construction) to be used:

Table 5.1

S.No.	Rated voltage of lamp V	Current rating A	Emf of one cell V	Number of cells N
1	1.1	–	1.5	1
2	2.2	0.2	1.5	2
3	2.5	0.3	1.5	2
4	3.8	0.5	1.5	3
5	4.2	0.5	1.5	4
6	6.2	0.5	1.5	6

From Table 5.1 it can be noted that multiplication of emf of one cell and numbers is higher than rated voltage of lamp. This is due to the fact that voltage of cell drops when load of lamp is applied, that is, lamp is connected to cell. This dropped voltage, on load, is called PD (potential difference). Drop of voltage depends on how much current a lamp is drawing and how much chemicals are yet active to produce electrical energy. Drop of voltage on load (PD) with a particular lamp would be less when a battery/cell is new. PD would start getting dropped with prolonged use of electrical energy of cell/battery. A lamp having a rated voltage of, say 2.5 volts, may draw a current of 0.3 amperes from battery. Another lamp having rated voltage of 2.5 volts, can draw more than 0.3 amperes, say 0.5 or 0.6 amperes. This may happen due to the difference of internal construction of lamp. Suppose a lamp is in short circuit condition from inside due to defective assembly, a huge amount of current from cell/battery would flow. This would be the maximum current which chemical reaction inside the cell could provide.

In this process of 'forced' generation of current, vigorous chemical reaction generates heat and generation of electricity gets collapsed soon. For this reason, lamps are also rated for current which is drawn by them at rated voltage. Current rating of various lamps is also shown in Table 5.1. Number of hours for which it produces light before failing is called life of lamp which is on rated voltage. In actual practice, life appears to be too long when light gadget is operated by batteries. This is due to the fact that PD of battery keeps on getting reduced with usage. Since life of miniature lamp is rated at fixed voltage, therefore for testing the life of lamp such a source of electrical supply is needed whose potential difference does not drop. A power supply is normally used where there is no PD drop. Hence life of lamp can be checked for rated life without its prolongation as it happens with dry cell/battery.

General Construction of Lamp

Figure 5.1 shows general construction of two miniature lamps.

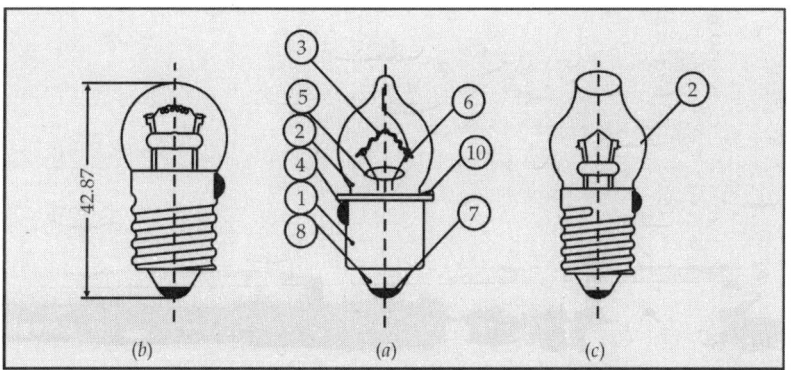

Fig. 5.1: General Construction of Lamps

Figure 5.1 (*a*) is of a pre-focused lamp. It is used on such light gadget where focus spot is fixed and cannot be varied. Figure 5.1 (*b*) is of a lamp which can be used on devices which have threads to receive lamp. These devices may or may not be of focusing type. Following are the parts of lamp:

1. Metallic cap
2. Glass shell/bulb
3. Filament
4. Solder
5. Electrodes
6. Solder terminal
7. Vitrite glass

Diameter of collar of cap of a pre-focused lamp is critical as it has to sit properly in a round seat of device in which lamp is to be assembled. Overall length and diameter are also important. In focusing type metallic cap, there are threads. Normally profile, minor and major diameters of cap are maintained as per BIS recommendations. This ensures proper fitting in the device as device makers also generally follow BIS.

Glass shell of miniature lamp performs three main functions, namely

- To contain other components
- To sustain high atmospheric pressure as there is a high vacuum inside assembled shell
- To allow maximum light to pass through it
- To provide desired shape and size to lamp

Glass shell ② in Fig. 5.1 (*a*) has a shape to hold internal parts ③, ⑤ and ⑥. Conical shape at the top of shell is for locating one of the electrode ⑤. In Fig. 5.1 (*b*), shell is spherical and holds other parts of lamp. Normally shells are produced from clear lead glass tubings, coloured shells are also produced to give coloured light.

Shell ② as shown in Fig. 5.1 (*c*) has a speciality that it has a formed lens at the top to provide a spot of light.

Filament

Filament is a light producing component. It is a very thin tungsten wire of only a few microns in diameter. Diameter of wire is so thin that it is difficult to measure by normal micrometer. Therefore the size of wire is expressed in milligrams per 200 millimetre length. Reason for making a filament with tungsten is that it melts or evaporates at a very high temperature of around 3200 degrees centigrade. At elevated temperature filament emits energy in the form of waves of various frequencies. Energy frequency in visible range produces a visible light.

Light, as seen, may be a dull light or a bright one. So to produce bright light, elevated temperatures are attained in filaments. If filament is heated in air, it will immediately start getting oxidized due to presence of oxygen in air. Filament can also break due to temperature rising to melting point. Design of filament is made in such a way that it emits bright light before temperature of filament reaches melting point. Many times it is observed that a lamp gets fused moments after lighting. This may be due to application of voltage higher than specified. For example, a 2.5 volt lamp is connected to a six volt battery. Figure 5.2 shows a number of filaments. Figure 5.2 (*a*) is just a straight filament wire. Figure 5.2 (*b*) is a straight filament wire having an inverted 'V' form shape. Figure 5.2 (*c*) shows a filament in the form of coil. That means straight filament wire is wound in the form of coil. Figure 5.2 (*d*) shows a filament coil which is bent in the form of bow. Figure 5.2 (*e*) is a coiled filament which is given a shape of inverted 'V'.

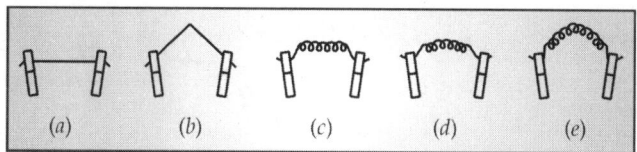

Fig. 5.2: Filaments

Filament Design

Fundamental principle of filament design is that it should emit maximum light for a fixed input of electrical power in watts. First of all designer decides the voltage on which lamp is to be operated and the value of current which filament of miniature lamp would draw. Following formula may be used to determine resistance of filament wire:

$$E = I \times R$$

where E is applied voltage, I is current to be drawn by filament and R is to be determined

$$R = E/I$$
$$= 2.5/0.3$$
$$= 8.33 \text{ ohms}$$

when E is in volts and I in amperes

A filament can be of 8.33 ohms with a number of combinations of diameter and length of filament wire. Diameter of filament wire may vary from 3.16 mg/200 mm length to 3.42 mg/200 mm length of wire (as mentioned earlier, this is practical way of indicating diameter of filament wire). Let there be three wires of following thickness (diameter).

3.20 mg/200 mm length of wire
3.22 mg/200 mm length of wire
3.30 mg/200 mm length of wire

3.20 mg/200 mm length of wire would be thinner as compared to 3.22 mg/200 mm length of wire. If specific resistance of wire per unit length is known then length of wire may be calculated to have a resistance of 8.33 ohms. In practice some adjustment in length is required because of change in resistance value when filament gets heated. Let calculated length of wire comes out to be,

for   3.28 mg/200 mm length of wire ------------ 5.36 to 6.19 mm
        3.30 mg/200 mm length of wire ------------ 6.86 to 8.01 mm
        3.33 mg/200 mm length of wire ------------ 7.62 to 9.15 mm

Now the question is which size of wire should be used so that current drawn with applied voltage (2.5 volts) does not exceed specified current (0.3 amperes) rating. Answer to this question may be found after further reading following para.

Normally straight length of tungsten wire cannot be accommodated between two electrodes of lamp. Therefore, to accommodate full length of wire, it is coiled with some straight length as shown in Fig. 5.3

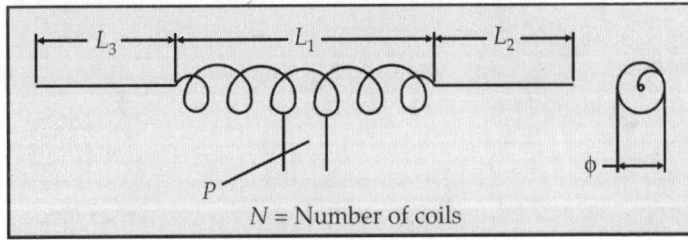

Fig. 5.3: Coiled Filament

It can be appreciated that there are following variables for designing a filament:

L_1 = Length of coiled portion of filament
ϕ = Diameter of coil
N = Number of coils

P = Pitch of coils
L_2 = Straight portion
L_3 = Straight portion

A number of combinations can be made from a single length of filament wire. Few possible combinations are shown in Table 5.2.

Table 5.2

S.No.	Wire size mg/ 200 mm	Coil length mm L_1	Coil diameter mm	No. of coils	Straight portion $L_2 + L_3$	Total length of wire in coil mm	Grand total Length mm	Pitch of coil mm
1	3.2	0.8	0.122	14	1.3	5.36	6.66	0.05
	"	"	"	14.5	"	5.55	6.85	"
	"	0.9	"	15	" ·	5.74	7.04	"
	"	"	"	16	"	6.13	7.43	"
	"	"	0.116	14	1.5	5.09	6.59	0.06
	"	"	"	15	"	5.46	6.96	"
	"	1.00	0.116	16	"	5.83	7.33	"
	"	1.10	"	17	"	6.19	7.69	"
2	3.30	0.9	0.122	14	"	5.36	6.86	"
	"	"	"	15	"	5.75	7.25	"
	"	"	"	16	"	6.12	7.62	"
	"	"	"	17	"	6.51	8.01	"
	"	"	"	14	"	5.36	6.86	"
	"	"	"	15	"	"	"	"
	"	1.0	"	16	"	6.12	7.62	0.07
	"	"	"	17	"	6.51	8.01	"
3	3.33	1.3	0.13	15	"	6.12	7.62	0.08
	"	"	"	16	"	6.53	8.03	"
	"	"	"	17	"	6.94	8.44	"
	"	"	"	18	"	7.35	8.85	"
	"	"	"	16	1.4	6.94	8.34	"
	"	"	"	17	"	"	"	"
	"	"	"	18	"	"	"	"
	"	"	"	19	"	7.75	9.15	"

Table 5.2 shows about 24 possible combinations. There may be still much more. There are two ways of selecting a combination for putting it in use. One method is by carrying out theoretical

calculations. Determination of a correct filament by theoretical designing is quite tedious and can be done by a theoretical designer. It is beyond the scope of this book to present theoretical designing. Other method is to select a few combinations from table. Take some quantity of filaments on experimental jig and machines. Load the filament in a small testing container. This container should have facility of getting its inside air evacuated by suitable vacuum pump. A minimum vacuum of 10^{-2} Torr is sufficient to carry out experiment. Trial equipment should also have DC voltmeter and ammeter to read applied voltage and current drawn by filament. There should be a photoelectric cell to have a comparative idea of light output. On trying out about ten combinations, three or four may be selected for producing trial lamps on regular production machinery, this is necessary to check the correctness of selected filaments. Another important aspect of quality of lamp is its life in hours when operated at specified voltage and drawing specified current. There are mainly three basic reasons on which life of filament depends. These are following:

- Thickness of wire
- Quality of vacuum
- Use of chemical called getter and its quality

By trial and error method such a design of filament is selected which has thickest (diameter) filament wire without over shooting specified current. Such a selected filament takes longer time to evaporate or melt, thus increasing life of filament in lamp.

Lamp's Efficiency

Many a times it is found that in spite of best selection, life of lamp remains less than expected. Although glass sealed shell has high vacuum inside, yet there is some residual oxygen which oxidizes filament wire to some extent, thus marginally reducing its life. Apart from this, possible impurities present in or around tungsten filament get evaporated and marginally reduce vacuum inside sealed shell. To overcome this deficiency, a chemical, known as 'getter' is applied to filament before it is assembled in glass shell for sealing and evacuation. Getter improves life of filament at rated voltage and current.

Large coil diameter tends to make coils flabby, hence there is a possibility of one or two coils touching each other, causing a short circuit. Filament of such type of shortcoming is not reliable. It may fail even during use of lamp. Coil of small diameter would necessarily have more number of coils thus increasing the length of coil to such an extent that straight portion of filament wire between first coil and electrode becomes too short. Too much nearness of coil to electrode causes loss of some heat from heated coil. This adversely affects the efficiency of filament. While selecting a design of filament, above mentioned points may be taken into consideration.

Pitch of coil is another variable to be considered while selecting a design of filament. Coils having less value of pitch may lead to mainly two defects. First is short=circuiting of coils. Secondly, longer straight wires L_2 and L_3 which normally do not contribute in generation of light, hence efficiency of lamp drops.

Practically, sometimes trials are repeated many times till best possible design could be arrived at.

While production is going on, quality checking also goes on simultaneously. Many a times very slight variation in L_2 plus L_3 is done to maintain correct light output with slightly increased drawn current, say 5 to 8% plus or minus.

Electrodes

Electrodes ⑤ and ⑥ are shown in Fig. 5.1. These are the wires which carry electric current from terminals ④ and ⑦ of lamp to filament ③. Wire is specially produced to have following features.

- Bends easily with least spring back
- Embeds in glass
- Has good electrical conductivity

Figure 5.4 shows construction of electrode wire.

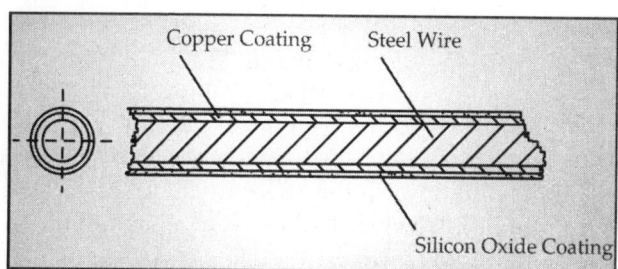

Fig. 5.4: Electrode Wire

Hook Formation

Main supporting wire is drawn to a diameter of 0.25 to 0.30 mm. Ductility of drawn wire is maintained to such a level that it does not spring back if a 180 degrees bent portion is pressed as shown in Fig. 5.5.

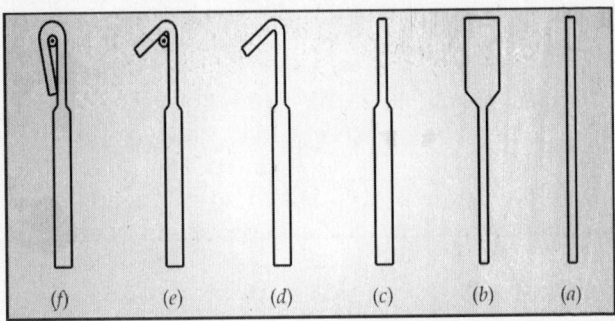

Fig. 5.5: Hook Formation

Figure 5.5 (*a*) shows straight wire unrolled from reel. Wire is generally supplied wound on a suitable diameter reel. Figure 5.5 (*b*) shows a portion of wire pressed to a flatness. Flattened length of wire is bent at about 45 degrees so that bent portion is about one third of flattened length. Hook so formed is meant to receive straight portion of filament. Bent flat portion is then pressed to such an extent that filament wire gets embedded in steel wire together with copper coating, thus creating a firm contact between filament tungsten wire and electrode.

Wire described above is generally called dumet wire. Since steel is not very good conductor of electrical current, especially at low voltage, it is coated with a copper film for better conductivity of electrical current and at the same time protect ferrous surface of wire getting rusted. But copper coating has two main disadvantages. Firstly, its surface gets oxidized as it remains exposed to atmosphere. Secondly, its sealing in glass mass is not perfect from vacuum tightness point of view. It is for these two reasons that copper coating is further coated by silicon oxide which is clear to transmission of light, therefore, copper coating can be visually examined. Moreover, adhesion of silicon oxide to copper and glass is good. Therefore, sealing of electrodes passing through glass is good against vacuum leak.

Soldered Terminals

One electrode is internally brought down to eyelet of metal cap and other electrode, to the slot at the brim of cap. This is shown in Fig. 5.6. ⑥ of Fig. 5.1 is one electrode which is protruding from slot in metal cap. Other electrode ⑤ is passing through brass sheet eyelet ③ which is fixed to cap by insert moulding of vitrite glass. In this way eyelet and cap are electrically disconnected completely. Excessive protrusions of electrodes are cut. Only a little portion, say 0.8 to 1.2 mm may protrude. Now soft solder is applied to mechanically and electrically join one electrode to cap and other to eyelet of cap. This is shown in Fig. 5.6.

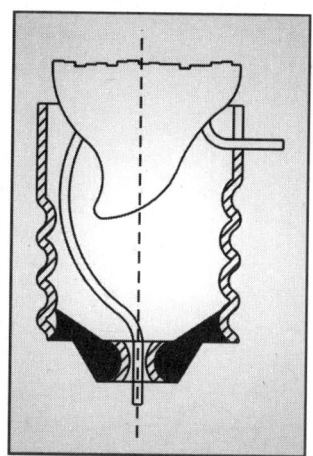 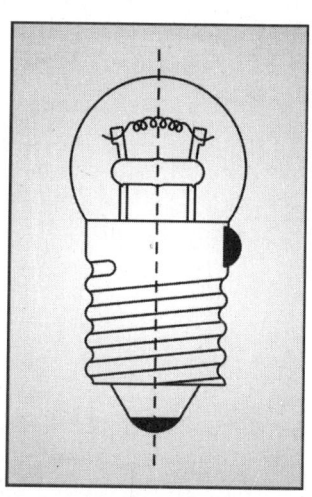

Fig. 5.6: Sealed Shell on Cap   **Fig. 5.7:** Soldered Terminals

Depending on assembly process, solder wires of various compositions of lead and tin are used. Generally solder wires of following compositions are used including negligible impurities.

Lead	Tin
70%	30%
75%	25%
80%	20%

Solder wire with flux filled in wire capillary throughout length of wire is also used. Selection of solder wire is normally done by trial and error method by producers of miniature lamps.

Vitrite Glass

Vitrite glass is used to join eyelet and body of a cap of miniature lamp. Eyelet and cap are mechanically joined. Electrically they are completely separated. In appearance, vitrite glass is dark in colour and opaque. Its composition is such that it does not crack when subjected to forming pressure during almost molten condition and cooling down to ambient temperature fast.

Dimensional Quality

Dimensional quality of miniature lamp is of importance because lamp should fit properly in lighting device, such as a flashlight. Good quality focus of a flashlight can be obtained if dimensions conform to specifications as recommended by BIS. Significance of various dimensions of miniature lamp is described with the help of Fig. 5.8.

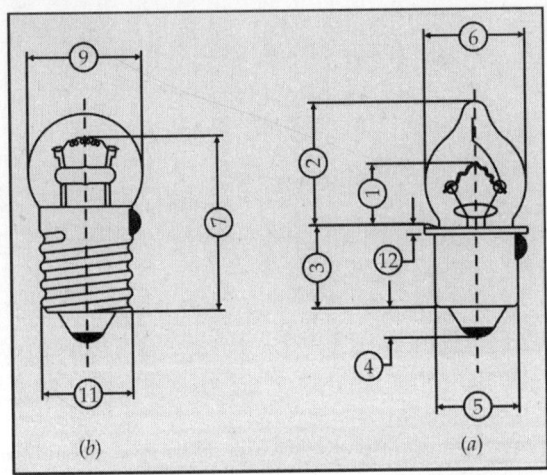

Fig. 5.8: Lamp Dimensions

Figure 5.8 (*a*) shows a pre-focused miniature lamp. While assembling in a flashlight, collar ⑫ sits in a bulb holder which is fixed to metal or plastic reflector. Good spot of focus from flashlight is obtained when filament lies on focal point of parabolic profile of reflecting surface. This can happen if height ① is maintained as per design within permissible limits. ② is not a critical dimension. It should not exceed higher limit because then there is a chance that tip of shell strikes lens of flashlight which is normally fitted. Dimensions ③, ④ and ⑤ are also not critical. Diameter of collar ⑫ is critical as it has to precisely fit in bulb holder recess. Diameter ⑥ of glass shell should not exceed higher permissible limit because then there is a chance that it does not enter the hole of bulb holder. A very important dimensional aspect is that centre of filament should lie on axis of cap within permissible limits of outness. This is important in the interest of achieving good focus.

Figure 5.8 (*b*) shows a focusing type miniature lamp. Diameter ⑪, profile of thread, minor diameter of threads are very important from assembly point of view. BIS specifies thread profile and related dimensions. Dimension ⑦ is important but not critical from focusing point of view. It is not designed to give a preset focus. Screw type lamps may be used in flashlights where focusing facility is provided. It may also be used for any type of lighting device where it is immaterial if light is focused or not. Diameter of glass shell should not exceed its permissible higher limit as it may have to pass through an opening of reflector or any other component. In this spherical lamp also, position of filament is very important if it has to be used for a device from which a good spot focus is to be achieved. Dimension ⑦ should be in specified limits. Moreover, filament centre should lie on axis of cap within permissible limits of outness.

In lens end type lamp formation of lens end on the tip of shell should be such that it gives a spot of light with distinct outline, not hazy brightness. Light in the spot should be almost uniform. Light should be white, that is free from refracted light giving tinge of rainbow colours. Dimensions of cap, shell, etc. are of normal importance.

Production Process

Miniature lamps can be produced by manual operations and by automatic machinery and plant. Manual process is adopted by tiny industries, catering to low-priced market slot. These lamps normally do not conform to standards laid down by BIS. Following is the sequence of process generally adopted for manual process:

- Blowing of glass shells
- Cutting of glass shells from tubes
- Cutting of required length of electrodes on jig

- Fusing of glass bead around electrodes
- Making of hooks manually by means of a jig
- Placing a filament in the hooks with the help of a tweezers
- Closing the hooks with the help of jig
- Hammering the hooks so much that tungsten wire of filament gets embedded
- Mounted beads are now placed inside glass shell
- Shell with mounted bead is placed over a jig of single head sealing machine
- Shell is sealed with a glass tube by flame tips. Vacuum is then created. Sealed shell is cut off by flame from glass tube.
- Sealed shell produced are tested for quality of vacuum by means of an ultra high voltage spark produced by a small instrument called 'Tesla coil'
- Caps are manually filled by cement and sealed shells are placed in it. Combination is then heated to such an extent that thermosetting cement is cured. Consequently cap and sealed shells are joined together.
- Protruded wires are cut to such an extent that a small length is left behind
- Soldering of wires is done

Tiny lamp producing parties procure mouth blown shells from glass craftsmen or glass blowers. Glass blowers hold a glass tube over the heat. Once opening of glass tube is pliable, it is closed with a small tap. Closed end glass tube is then rotated by hand. When glass tube near closed end gets sufficiently heated, a balloon is mouth blown. It is now the shell which is parted out from tube by blower. Shells so produced are very clear and smooth. But there is always variation in diameter and length of neck. Shell blowers supply shells to parties after sorting visually.

Electrode wires are cut to pieces of required length by a hand operated cutting jig. Pairs of cut wires are loaded on another jig as shown in Fig. 5.9.

Figure 5.9 (*a*) shows the electrode pieces placed inside the holes. Depth of holes is such that required length of electrode pieces remains above the surface of jig. Figure 5.9 (*b*) shows a glass bead placed around pair of electrodes. Figure 5.9 (*c*) shows that with the help of a burner flame glass is fused around and between the pair of wires. This is done by manipulating position of flame around glass beads. Generally amorphous glass beads are used and not sintered glass beads. Figure 5.9 (*d*) shows a mounting which is removed from jig with the help of a tweezers.

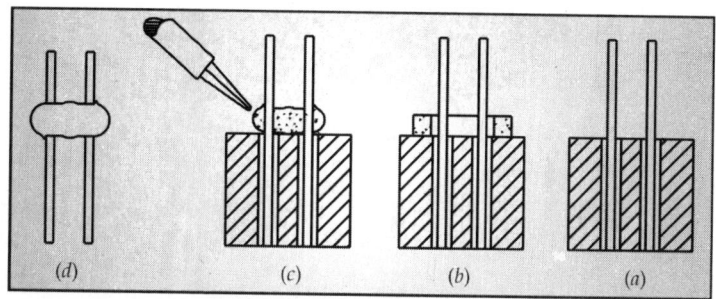

Fig. 5.9: Fusing of Beads

Next step is formation of hooks on both electrodes. Basic design of a jig is shown in Fig. 5.10.

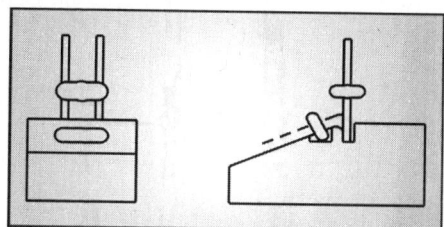

Fig. 5.10: Hook Formation

Portion of electrode with bead outside the jig is pushed manually to approximately 45 degrees. This action of worker forms the hook. Bead mounted electrodes are now ready to receive filament.

Operation of filament loading is carried out on next jig. It is shown in Fig. 5.11.

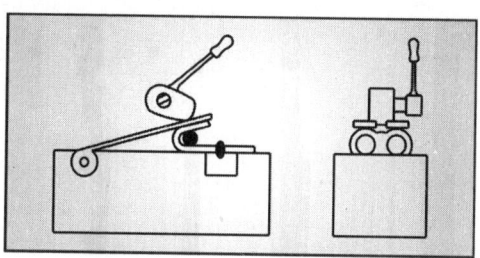

Fig. 5.11: Filament Loading

Bead mounting is placed in between the jaws. Lower jaw is the body of jig and upper jaw closes by pulling the handle. Design of pressing cam and length of handle is such that open hooks of electrodes are first get bent to 180 degrees with respect to electrode axis. Further pulling of handle presses the hooks so much that straight portion of filament wire gets embedded in electrode. Now mounted beads are collected in a suitable plastics shallow container. This container is then shifted to a single head sealing machine. Figure 5.12 shows basic construction of a single head sealing machine.

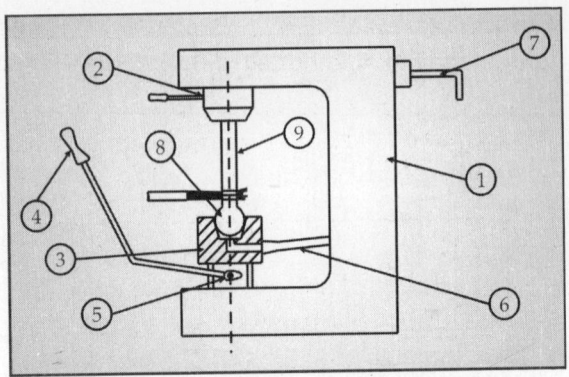

Fig. 5.12: Single Head Sealing Machine

① is frame of single head sealing machine. ② is a chuck to hold a exhaust tube. ③ is a shell holding seat which has a vacuum connection ⑥ to hold shells ⑧ in place with mounted bead already placed inside the shell. There are three or four burners to heat the edges of exhaust tube and shells simultaneously. Set of burner mounting can be swivelled in and out.

When proper temperature of edges of exhaust tube and shell is attained as judged by worker, lever ④ is pulled to raise shell so that red hot edge of shell is slightly pressed so that electrodes get embedded into soft glass and a joint is established between shell edge and exhaust tube edges. When glass of two parts get fused in each other then this joint is allowed to cool down and heating of exhaust tube a little above the joint is initiated again. Once exhaust tube is sufficiently heated, lever ④ is released slowly so that shell holding seat ③ comes down with shell. By doing so a capillary is formed at the heated portion of exhaust tube. Vacuum line ⑦ is switched on. By doing so evacuation of shell starts. After a little time, say half a minute, capillary is heated again so that it closes and cuts from exhaust tube. In this way sealed shell is obtained and collected. Occasionally quality of vacuum is tested to ensure that process is going on correctly.

Sealed shells are brought to another table where a worker assembles sealed shells on cement filled caps. Caps with sealed shells are placed in a heated die and allowed some time, say half a minute for cement to cure. Caps with shells are carefully removed to cool down so that cement becomes hard and its grip over glass of sealed shell and inside of cap becomes strong. Now the lamp is ready for hand marking and packing.

Production by Automatic Machines

Description of manual process of producing miniature lamps indicates that there are a number of operations carried out manually. Inherent shortcomings of many manual operations are undesirable variation in quality, higher rejection, poor productivity, higher production cost and more IR (industrial relation) problems. For all these reasons, medium and large industries usually opt for automatic plant and machinery.

A number of automatic machineries make a complete lamp making plant. Some machines are fed manually by input material or components. Output of one or two machines is the input for other machines. If transfer of output of one machine to another machine is mechanized then the plant may be called a fully automatic plant. First of all basic working system of following machines is described:

- Bulb blowing machine
- Bead mount mill
- Sealing machine
- Cap filling machine
- Capping mill
- Quality testing laboratory

Bulb Blowing Machine

Basic construction of machine is described with the help of Fig. 5.13 (*a, b, c, d, e* and *f*).

Fig. 5.13 (a): Arrangement of Stations

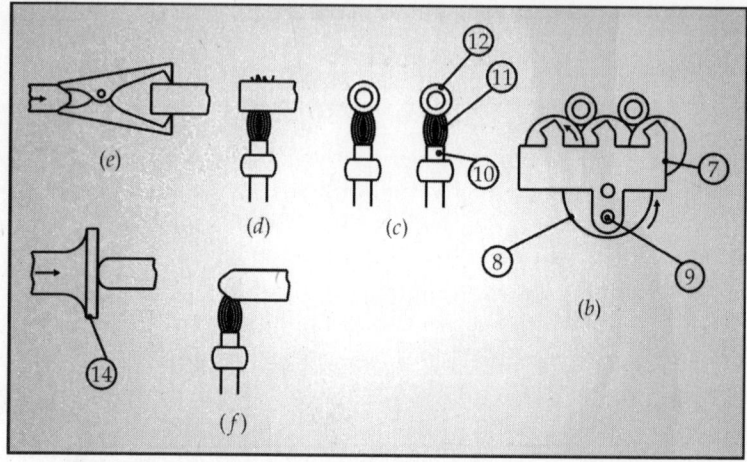

Fig. 5.13 (b, c, d, e, f): Basic Construction of Machine

Figure 5.13 (a) shows general arrangement of various machine parts which are handling lead glass tubing. ① is a chute where lengths of glass tubings are stacked. There is a tube release system which releases only one tube at each stroke of tube transfer mechanism. ② are rotating wheels on which tubes rest in such a fashion that two wheels rotating in clockwise or anticlockwise direction make tubes continuously rotate except when under transfer from one position to another. ③ are rotating glass tubes. ④ is a jet of very cold air (chilled). ⑥ is a hard steel disc having a knife edge sharpness throughout its circumference. Air jet ④ keeps the sharp edge of wheel cool. This arrangement is at last but one station. There are small burners ⑩ under each rotating tube. There are a number of stations of rotating tubes. Tubes are transferred simultaneously from one station to next by means of pair of long transfer plate ⑦. Motion to plates ⑦ is provided by an eccentric pin ⑨ fitted on a rotating disc. So the motion of plate ⑦ is such that pushes all the tubes to next station. This process keeps on repeating continuously.

Tips of glass tubes over burners are heated by varied quality of flames. First flame is a flabby flame of a mixture of gas (say LPG) and air. At this station glass tube is warmed up. On next station flame quality is such that glass becomes visibly red hot. On next station it is further heated to such a softness that pinch off of hot tip is possible on next station. After pinch off, closed end of tube is still heated. On next station pinched off and closed end of tube is pushed a little by a pusher ⑭ to make the hot end flat. On next station a reducing flame heats the tip to an almost clear hot glass tube end. Just after this station is blowing mould. As soon as hot glass tube end reaches between two halves of bulb blowing mould, mould gets closed and high pressure air is injected from the open end of tube. Shell is instantly blown while it is rotating in the mould. As soon as mould opens, glass tube with blown shell is transferred to next and final station. At this station, temperature near the neck of shell is maintained to some value (not visible heat) to suit cutting/parting off on next station. As soon as tube with blown shell reaches cutting station, cold sharp wheel comes down to touch hot surface of glass tube. Due to thermal shock, shell gets perfectly parted off in a straight plain at right angle to tube axis. After parting off, shells are collected in a container. Sometimes shells are placed in an oven at a temperature of about 250 degrees centigrade for about half an hour for removing internal stresses in glass shells. Shells are cooled down slowly to ambient temperature. Now the shells are ready for manual transfer to bead mount mill and placed near the input conveyor of bead mount mill.

Bead Mount Mill

Basic working principle of bead mount mill, as it is called, is explained with the help of Fig. 5.14.

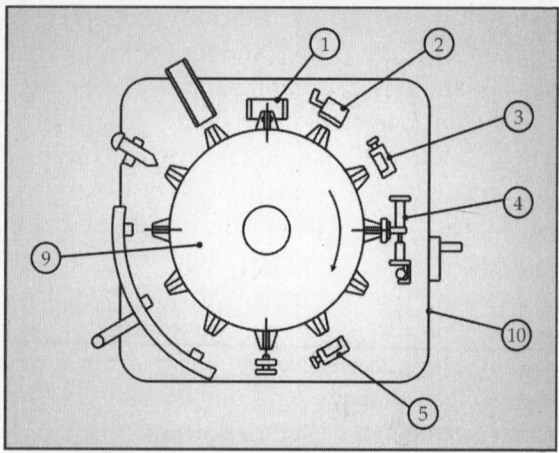

Fig. 5.14: Basic Bead Mount Mill

⑩ is a platform of cast iron on a four-legged stand. Height of platform is suitable for a person to work in standing posture. Sides of square platform may be around one metre. Machine has an intermittently rotating turret ⑨ fitted in the centre. Turret is fitted with fourteen or more fingers which are designed and made to hold electrode wires which are fed at station ① and cut before turret indexes. Station ① has an oven through which pair of electrode wires pass. There are two reels of wires underneath the oven under table ⑩ of machine. Since wire unwinds from reels, have curvature, it is necessary to make the wire straight. This happens when wire, under tension, is heated in the oven. Turret finger holding cut electrodes may be called F_1 for the purpose of ease in explaining progressive operations carried out at various stations of BML (bead mount mill). F_1 is indexed to station ② where sintered glass bead is fed around electrodes. On next station, gap between two wires above F_1 is increased by a wedge. Increased gap is maintained to have designed dimensions. Turret again indexes to bring F_1 at station ③. At this station, electrode wires partly flattened to desired thickness and length. Turret indexes again to reach station ④. This station performs a number of complex operations, hence needs somewhat detailed description (Fig. 5.15).

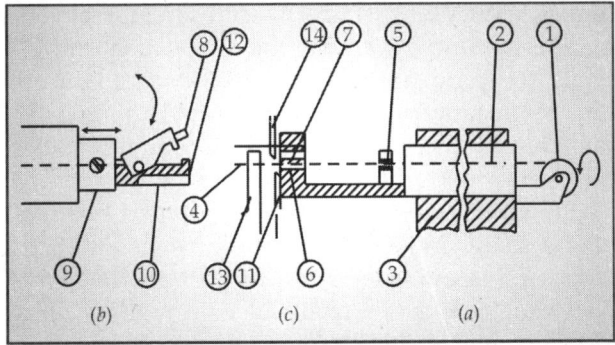

Fig. 5.15: Coiling Unit

Figure 5.15 (*a*) is a unit which carries filament wire reel ① and rotates with spindle ② in housing ③. There is a hole all along the intermittently rotating spindle. Filament wire ④ is threaded through spindle, through gripper ⑤ and fine hole ⑥ at the end of spindle. Wire is kept a few millimetres protruding outside the face of spindle. Spindle ② is mechanically attached to a driving gear box underneath machine table. Gear box also drives cam shafts which actuate various units, fitted over the table, with the help of turn buckle levers. Referring again to Fig. 5.15 (*a*), ⑦ is a hole to receive mandrel wire ⑧. Figure 5.15 (*b*) shows a reciprocating spindle ⑨. ⑩ are jaws. Upper jaw has a hole in which mandrel wire slides precisely. After a certain amount of travel of spindle ⑨, jaws ⑩ close over filament wire, thus holding the wire. At this time, gap between the face ⑪ of spindle ② and face ⑫ of jaw ⑩ is about one to two millimetres. Just after this stage, spindle ② starts rotating with high speed (about 300 rpm), with spindle ⑨ moving back with a speed of pitch of filament per round of spindle ②. As soon as number of coils are completed, cutter ⑭ gets activated and cuts filament wire. Now tweezers ⑬ come to hold filament while mandrel gets extracted from filament. Now tweezers carry filament to such a position that straight portions of filament get placed in hooks of electrodes. Immediately another pair of anvil press the hooks to close and press to such an extent that tungsten wire gets embedded in electrodes.

It is worth noting that so many critical operations are carried out by machine with precise sequence and timing. This is achieved by perfect design of machine elements and their assembly.

As soon as filament is loaded on electrodes, turret gets indexed to bring F_1 on station ⑤ where getter application unit is fitted. At this station lower portions of electrodes are given a specified formation. After station ⑤ there are a number of stations, three or four, which have burners to fuse bead around electrodes. Flames of these burners are so adjusted that complete and strong fusing of bead is achieved. F_1 then reaches at mounted bead unloading station and unloading takes place on a conveyor. On each indexing a mounted bead is unloaded from turret finger. Mounted beads are then manually picked up from output conveyor of bead mount mill and placed inside shells which are already prearranged in a tray of, say 100 shells positioning depressions. Operator of bead mount mill keeps on picking mounted beads and placing them inside shells. When full tray is completed, another tray is picked up by operator. Filled trays are slided near input station of next machine. Operator, while placing mounted beads in shell, also has an eye on functioning of bead mount mill.

Vacuum-cum-Sealing Machine

Function of sealing machine is to join glass exhaust tube to shell together with mounted bead, to draw a capillary, to create vacuum in shells, to seal the capillary, cut out sealed shells from exhaust tube over an outlet chute and remove used exhaust tubes.

Sealing machine comprises two units:

- Exhaust tube and shell joining and capillary formation unit, Let us call it unit 1.
- Vacuum creating, sealing and sealed shell parting off. Let us call it unit 2.

    Both the units are intermittently rotating turrets. Unit 1 has about 18 to 24 heads equidistant from centre of turret and also at precisely equal intervals on PCD (pitch circle diameter). There are following stations:
- Manual placement of shells on turret heads
- Automatic feeding of exhaust tubes
- Heating and fusing of exhaust tube end to glass shell
- Drawing of capillary
- Transfer from one turret to another

Direction of indexing of this unit is usually clockwise when seeing the turret from top.

Unit 2 also has an intermittently rotating turret. Its direction of rotation is anticlockwise. Sealing machine is briefly explained with the help of Fig. 5.16.

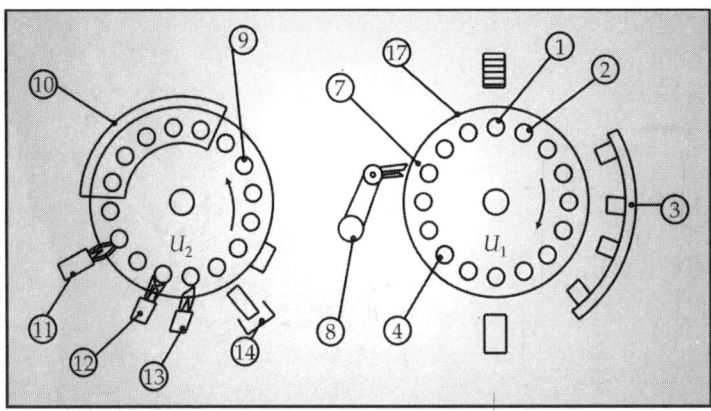

Fig. 5.16: Turrets

Unit 1 and 2 of sealing machine are mechanically synchronized in such a way that turret of unit 1 intermittently rotates in clockwise direction and turret of unit 2 in anticlockwise direction. Both the units have, say 18 heads. ⑰ is the station where shells with mounted beads are manually fed to heads of turret. Shell is mechanically held in position as soon as turret moves. ① is the station where exhaust tubes are stacked in a hopper. Just after completion of each indexing, one exhaust tube is automatically fed to a head which has just stopped in front of station ①. Exhaust tubes are fed to heads in vertical position in such a way that lower end surface of exhaust tube is about two millimetres above the surface of shell neck. While turret is getting indexed, shell neck and exhaust tube end pass in front of a number of burners (please note that heads are now rotating). These burners have flames so adjusted that glass is first warmed, heated to visible heat and then to a fusing heat. At this station, head is slightly lifted by means of under the table follower cam. This lifting fuses exhaust tube to shell neck. On next few stations ④ exhaust tube is heated to pliable condition and head is pulled down on station. Consequently, a capillary is formed to such an extent that it is strong enough to sustain transfer and vertical push on coming station. Now heads stop rotating. Station ⑦ is the station where exhaust tube with shell is transferred from unit 1 to unit 2. Transferring action is carried out by a mechanically synchronized arm ⑧. Transferring takes place when both the turrets are in stationary mode (not indexing). Arm has a swinging vacuum gripper which grips exhaust tube. It moves back only when grip of finger of head on turret, unit 1, gets loosened. Arm brings exhaust tube just under the head ⑨ of turret of unit 2. Immediately a pusher from below pushes exhaust tube in the head, whose hole is loose for exhaust tube to get in completely to required length, grip of rubber seal gets tightened to make a vacuumtight gripping. All the heads pass through heating oven ⑩. Head is connected to vacuum system as it enters oven. Heat from oven facilitates creation of good vacuum. All the heads from the position of oven, till sealing and cutting off, are connected to high vacuum system of around 10^{-5} torr or better. As soon as an exhaust tube with shell comes out of oven, it is immediately in front of a flame which increases the temperature of capillary near the joint of shell and exhaust tube. Next station has a fierce flame ⑬. As soon as exhaust tube cum shell reaches, capillary gets collapsed to complete closure and glass melts to such an extent that sealed shell falls down in an outlet chute. There is also a 'tesla coil' tester ⑫ fitted to keep on visually indicating quality of vacuum. Sealed shells are collected in a small bin ⑭ which is manually transferred to table where worker 'threads' the electrode through eyelet and slot of metal cap and presses the shell on cap filled with cement and then places in another tray having 100 depressions for placing shells with caps, caps on top and shells in depressions. These trays are manually transferred to input station of capping mill.

Cap Filling Machine

Cement, which is generally used for joining sealed shell to metal cap, is a thermosetting type of cement. Lamp manufacturers normally procure cement ingredients from suppliers and knead the ingredient to a consistency like the flour kneaded for preparing breads. Kneaded cement is then placed inside the hopper. Function of automatic cap filling machine is described with the help of Fig. 5.17.

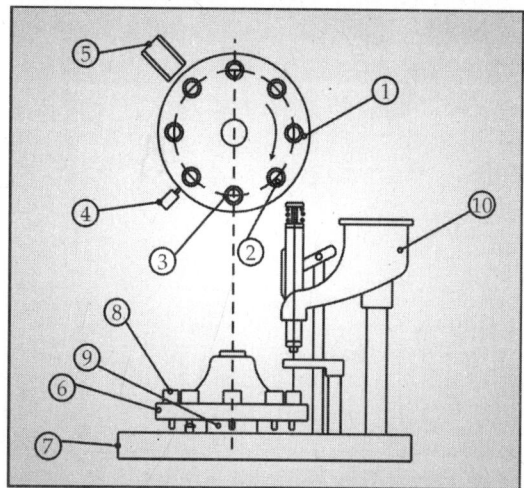

Fig. 5.17: Cap Filling Machine

Machine consists of following units:

- Machine stand ⑦
- Intermittently revolving turret ⑥
- Heads ③
- Turret reciprocating-cum-indexing spindle ⑨
- Manual feeding station ①
- Slot cutting unit ②
- Cement filling unit ⑩
- Marking unit ④
- Ejection station ⑤

On feeding station cap is manually placed by operator in the head. On indexing, head reaches slot making unit. Turret automatically gets lifted to a fixed height. Effective height of all units are made accordingly. Cap enters slot making punch and die. Unit operates to cut slot while turret is still in up position. Turret comes down and gets indexed. Now cap comes under cement filling station ⑩. Turret gets lifted. Consequently, cap pushes a flat sensor strip with a hole over to cement filling nozzle. Filling unit automatically operates and fills the cap with a set quantity of cement. In case there is no cap in the head, sensor strip would not be lifted to its set height. Therefore, cement filling unit plunger would not operate. Figure 5.18 shows progressive operations on cap.

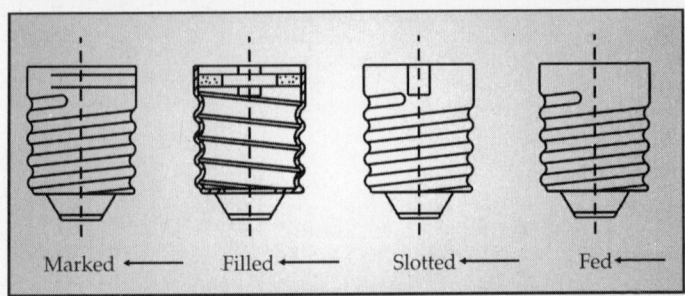

Marked ⟵——— Filled ⟵——— Slotted ⟵——— Fed ⟵———

Fig. 5.18: Cap Filling Machine Stations

One or two stations are blank. Head reaches next station which is marking station. As soon as turret rises, marking unit marks on circumferential surface of cap. After leaving one or two stations is the ejection station. On this station there is a pin. As soon as turret goes down, pin stops cap from going down. Hence ejection of cap takes place from head. In this way caps are continuously filled and ejected. Filled caps coming out of machine are collected in a box and are manually transferred to a table where sealed shells are threaded in caps. Trays are then manually transferred to capping machine.

It is worth mentioning that all the movements with designed and critically timed are carried out by mechanical synchronization.

Capping Mill

Basic construction and functioning of capping mill are described with the help of Fig. 5.19.

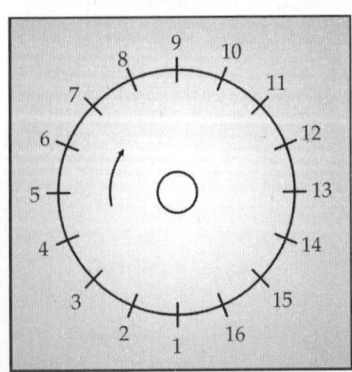

Fig. 5.19: Capping Mill Stations

Capping mill turret is of large diameter, say about one and a quarter metre. There are about thirty-four heads of design as shown in Fig. 5.20.

① is a cast iron head in which there is a hole with a seat on which face of cap rests. ② is a round block having a conical bore where glass shell rests. Round block ② is fitted on a spring loaded lever ③. At the loading station ① operator pulls down lever ③, inserts cap in the head and releases lever softly to hold shell plus cap in position. Head gets indexed to another

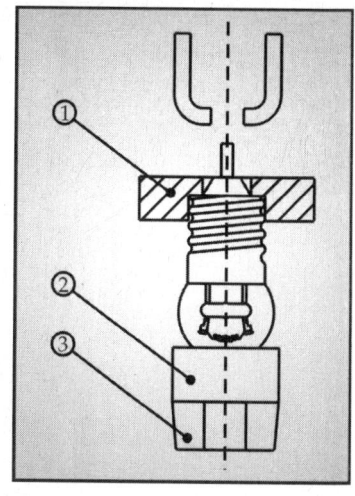

Fig. 5.20: Capping Mill Heads

station ② after two or three blank stations. At this station protruding wire (electrode) on top is pulled up with slight grip enough to make the wire protrusion erect. This action gets completed just before start of indexing. At next station excessive protruding wire on top is cut. On station ④, Fig. 5.19, top soldering unit applies solder on eyelet where a sharp flame is impinging. Flame is so set that good shape and strength of soldering is achieved. Head moves further to reach next station ⑤ after a gap of two stations. On station ⑤, side wire (electrode) gets stretched to make it straight. On next station excessive wire gets cut. Only about 1 mm remains protruding out of slot in the cap. On reaching station ⑥ soldering unit applies side soldering in presence of a sharp flame tip pointed exactly on protruded wire (electrode). Flame should be so set that desired shape and strength and electrical conductivity is achieved. Now the head travels intermittently in front of an array of burners, say ten to twelve which heat heads of turret in which caps with cement are there. Cement gets enough time to get cured and cool down to a reasonable temperature before reaching unloading station.

At unloading station two pins travel down. One of them is a little advanced so that lever is pushed down first and then lamp. A sliding chute is there on which lamp fells softly and slides down in a container. Now the lamps are ready for inspection before packing.

It is worth mentioning that it is possible to eliminate manual operations and introduce automation. One has to examine techno-commercial feasibility of automation.

Quality Checking and Control

Automatic miniature lamp making machines discussed so far produce lamps at a speed of about one thousand five hundred lamps per hour. If some defect crops up in any machine at any of its unit, a large number of defective lamps may get produced. This is a highly undesirable situation from economy and commercial point of view. Operators of various machines are trained to see and detect if some defect crops up. If it so happens, machine is stopped for readjustment. Stopping of machine disturbs the process, thus causing more rejection.

First step is quality checking. Some quality aspects may be visually checked by operator and quality inspector, like marking on cap or solder shape. Other quality aspects like current consumption, light output, life, etc. can be checked in quality checking laboratory which is equipped with instruments to check light output in lumens, voltage, current and lumens per watt. Life of lamps is checked at increased voltage. Results are pre-calibrated for real time for which lamp gives light.

Suppose expected life of a 2.5 volt-0.3 amps lamp is forty hours when lighted continuously. This is too long a time to take corrective measure if on testing life of lamp comes to be less than required. During this period thousands of lamps with less life would have been produced. So, to overcome this problem, testing of lamps for life is done on increased voltage, say 3.2 volts for a lamp of 2.5 volts. So lots of lamps under testing (say 50 lamps) get fused within two hours. Results of life testing on elevated voltage are corelated with life testing results at specified voltage by repeating experiments many times, doing statistical studies and preparing appropriate graphs. Light output on specified voltage is checked by lighting a lamp in a cubical or sphere which is painted white inside. There is a photoelectric cell. When light falls on a white screen in front of photoelectric cell, an emf is generated which passes through a resistance, current passing through a fixed resistance is measured by a digital microampere meter. Readings are calibrated with the help of lamps of known lumens.

In test laboratory there is a jig to test grip between glass shell and cap. Cap is held rigidly in the jig and torsional force is applied to sealed shell by a gripper. This indicates as to how strongly cement is holding glass shell and cap together.

Filament should be located on the axis of cap with tolerance of, say 0.2 mm radius. In first place, this is visually checked. Lamps with too much out filament can be visually identified. If necessary, sorting is carried out before packing of lamps in unit cartons is done.

Photograph

Photo 5.1

Lamp marked 1 is a prefocus lamp. Surface of collar towards glass shell is the reference surface. Lamp marked 2 is a focusing type. It is screwed in the focus adjusting system. Lamp marked 3 has a lens, formed in glass shell while it is blown.

6

Quality Assurance in Production Processes

Introduction

Present scenario of industrial products marketing in India and elsewhere is highly competitive. This is due to the fact that large section of consumers has become very quality conscious. It is still on rise due to presence of superior quality products from indigenous and foreign industrial houses. Any industry, medium or small, has to strive for following if it has to stay with firm footing in the market.

- Competitive pricing
- Excellent quality
- Consumer-friendly service

This chapter is devoted to quality assurance in production processes. Basic principles of quality assurance are dealt here with irrespective of any particular process. In fact, basic principles are generally applicable to most of production processes.

First of all it is necessary to define quality requirements of a product in terms of its design, material, dimensions, tolerances, colour, reliability, service life, aesthetics, etc.

Once quality requirements are defined, it is necessary to examine various plants and machinery because they have their own capabilities. Some plants and machinery perform various production operations with greater accuracy, whereas some other similar plants perform production operations with a variation beyond acceptable limits of dimensions and finish, etc. So, one of the important aspects of quality assurance is accuracy and reliability of plant and machinery. Many plants and machinery may work nicely in the beginning but after production for a few months, start producing defective quality. For example, a machine having a journal bearing would develop excessive running clearance as compared to ball or roller bearing in place of journal bearing. Another example may be of a steel die which wears out fast, thus producing defective components. A die made of tungsten carbide may last longer before start producing defective components.

Quality of material or components which are fed to machine are of paramount importance in ultimately producing excellent quality. An example of input component is described here with the help of Fig. 6.1 to highlight importance of good quality of inputs to achieve good quality of output.

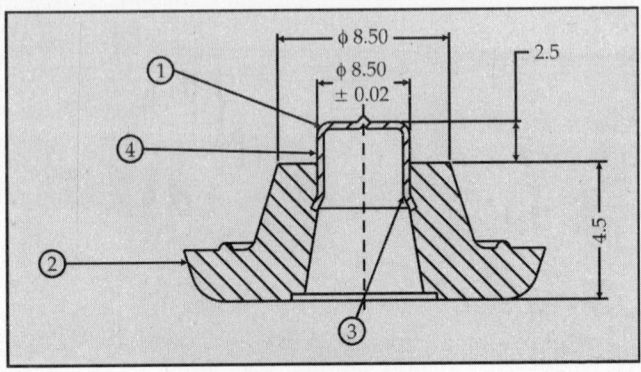

Fig. 6.1: Component

Component shown in Fig. 6.1 is insert moulded. ① is a sheet metal cap. ② is thermoplastic. ③ and ④ are those points where no flash is allowed. Gap between cap diameter and mould cavity bore should not exceed 0.03 mm. Clearance above this value will lead to development of flash of plastic. Bore in cavity where cap gets in, does not change within reasonable time of usage. It is the diameter and bore of the cap which may be attributed to formation of flash. It may be appreciated that how critical are the dimensions of cap which is an input to process.

For producing component shown in Fig. 6.1, thermoplastic is an input for insert moulding of component. Any variation in melt flow index beyond specified limit may cause formation of flashes. Hence it is highlighted that to ensure excellent quality of component shown in Fig. 6.1, quality of inputs should also be good.

Process Parameters

A production process has certain parameters to suit input material or components. Therefore, to achieve excellent quality of components, production process should also be under predetermined setting with tolerance. Following may be the parameters.

• Speed of process
• Temperature
• Accuracy in sequence of operation
• Time period of various operations
• Various forces and pressures
• Smoothness of motion, no jerks and vibration

Importance of speed of process can be appreciated with the example of blowing of PET bottle. After injection moulding of PET preform at injection station, core of the mould with pre-form still hot gets shifted to blowing station. In case speed of indexing slows down, it may not be possible to blow the pre-form fully to take shape of blow mould.

In another example, a plastic washer from the hopper slides down over the head of turret. If speed of turret increases, it may not be possible for washer to reach in time. It can, therefore, be concluded that speed of process is also important. In continuous process, such as extrusion of plastic filaments (thin strands), speed of process is very important.

In many processes, maintenance of correct temperature at various stages of operation is essential. In glass industry maintaining pre-determined temperature is necessary to give various shapes without development of internal strains. In processing of plastics, maintenance of temperature at various heating zones yields desirable quality product. There are innumerable examples where maintenance of correct temperature is highly desirable, for example, in sugar mills, milk packing plants and many chemical processes.

In production process there may be a number of operations which are to be carried out in a sequence. Sequences may be mechanized or built-in in the machine. There may be some manual actions also somewhere in sequential operations. Deviation from preset and predecided sequence of operations may render product completely unacceptable, that is bad or of defective quality.

Example of possibility of deviation in sequence of operations may be assembly of a few components of an electric switch. It is possible that worker puts that component in the die which has to be placed next to another component. Consequently, a defective switch is produced.

In fully automatic machine, like plastics injection moulding machines, timings of a number of operations are adjustable and programmable. After plastic is injected in mould cavity, some period of time is given (say five to ten seconds) for moulded component to solidify enough so that it retains its shape during ejection from mould cavity and core. If the timing is set less by mistake or due to any fault in the system, moulded component is ejected prematurely hence making it defective due to over shrinkage or even distortion.

In production process of chemical or paint industries there may be stages where specified period of time is given for completion of chemical reaction between various chemicals or additives. If this time gets increased or decreased, incomplete or defective reaction may be there, thus jeopardizing complete continuous process.

Certain chemical reactions do take place under specified pressures in reaction vessels. If the process is carried out under reduced or increased pressure, resultant output may not be up to the mark. Another example is of plastic blow moulded article which is blown in a mould by high pressure air. If the blowing pressure gets reduced, there is a possibility that fine details of cavity surface are not formed on the walls of blown article. In case of hot forging of parts, definite amount of force is required to forge part completely to desired shape. Any reduction in force would not complete the shape of compactness of forged part. Same holds good for coining operation as well.

Smoothness in Motion

In many delicate production processes, smoothness in motion of various machine elements is very important. Non-smooth motions give rise to vibration in machine or a particular unit of machine. Dynamic disbalancing of rotating or reciprocating parts also create vibration. Referring back to sealing machine (Chapter 5) on which shells with hanging mounted beads are placed on indexing head. If there is vibration in machine then mounted bead would shake.

Consequently, it may get sealed in glass when it is not in correct position. In such an event filament may be much out of axis.

Quality assurance of product can be established after the study of production processes. Such studies are undertaken under various conditions.

Case study

This aspect of quality assurance of product can be very well explained with the help of a typical case study.

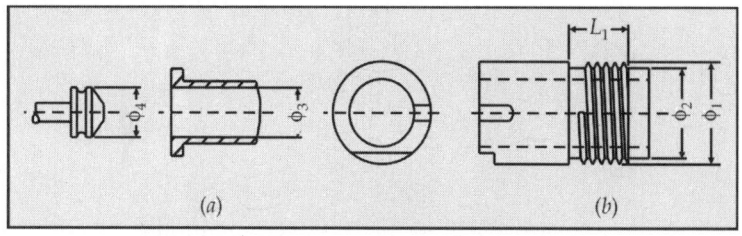

Fig. 6.2: Precision Plastic Components

Figure 6.2 (*b*) shows a threaded part of a household gadget. This part is moulded of polypropylene thermoplastic. It is moulded by a four cavity mould. Moulding cycle is six shots per minute, hence producing about eleven thousand pieces per eight hours shift. Supposing plant is working for 16 hours then production per day would be twenty-two thousand components. Component is to be precisely assembled in another component. Designed dimensions are

ϕ_1 = 24.50 mm ± 0.05

ϕ_2 = 22.20 mm ± 0.03

l = 12.50 mm ± 0.10

Since production speed is quite high, therefore, it is necessary to know at what time, under what circumstances dimensional deviation crosses specified tolerance limits, rendering a large number of components rejected. Once the pattern of deviation occurrence is known, corrective measures can be taken to eliminate undesirable deviations. First of all samples of component from running production are collected after every hour of production, starting from start of production shift. Let the sample size be ten numbers of components. This means that there would be eighty sample components per shift. All sample components should be measured and mean of readings be recorded in Table 6.1.

Date	Time	ϕ_1 24.50 ± 0.05	ϕ_2 22.20 ± 0.03	L 12.50 ± 0.1
18.08.2007	8.10 AM	24.55	22.23	12.55
	9 AM	24.50	22.20	12.55
	10 AM	24.50	22.17	12.50
	12 Noon	24.50	22.20	12.50
	1 PM	24.40	22.17	12.45
	2 "	24.45	22.15	12.40
	3 "	24.40	22.17	12.45
	4 "	24.50	22.20	12.55
	5 "	24.55	22.20	12.60
	6 "	24.50	22.23	12.60
	7 "	24.40	22.15	12.45
	8 "	24.40	22.15	12.50
	9 "	24.50	22.20	12.50
	10 "	24.50	22.20	12.50
	12 Midnight	24.55	22.23	12.55
	1 AM	24.55	22.20	12.50
	2 "	24.50	22.17	12.45
Min.	–	24.40	22.15	12.40
Max.	–	24.55	22.23	12.60
Avg.	–	24.47	22.19	12.50

Table 6.1 (title above)

Readings in Table 6.1 are now shown in Graph 6.1.

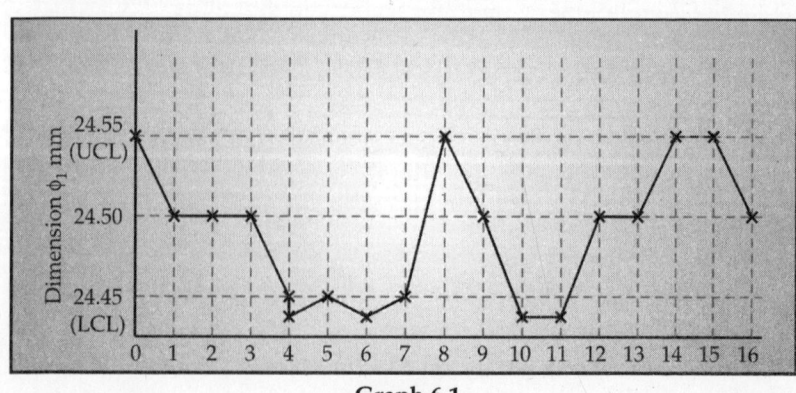

Graph 6.1

		Table 6.2		
Date	*Real Time*	ϕ_1 24.5 ± 0.05	ϕ_2 22.20 ± 0.03	*L* 12.50 ± 0.1
1.09.2007	8.10 AM	24.60	22.30	12.60
	9 ,,	24.60	22.30	12.60
	10 ,,	24.55	22.25	12.50
	11 ,,	24.50	22.23	12.50
	12 Midday	24.40	22.15	12.40
	1 PM	24.40	22.10	12.40
	2 ,,	24.45	22.17	12.40
	3 ,,	24.45	22.17	12.50
	4 ,,	24.50	22.20	12.50
	5 ,,	24.55	22.25	12.60
	6 ,,	24.55	22.23	12.60
	7 ,,	24.50	22.20	12.50
	8 ,,	24.40	22.10	12.40
	9 ,,	24.40	22.10	12.40
	10 ,,	24.35	22.05	12.40
	11 ,,	24.35	22.05	12.40
	12 Midnight	24.40	22.10	12.50
	1 AM	24.45	22.15	12.50
	2 ,,	24.50	22.20	12.50
Min.	–	24.35	22.05	12.40
Max.	–	24.60	22.30	12.60
Avg.		24.47	22.17	12.50

Readings in Table 6.2 are now shown in Graph 6.2.

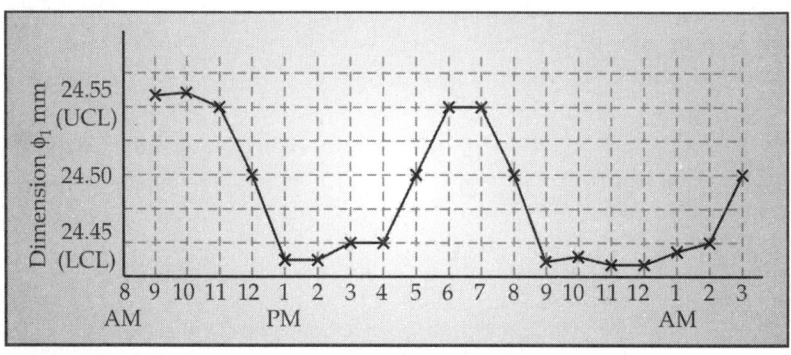

Graph 6.2

	Table 6.3			
Date	*Real time*	ϕ_1 24.50 ± 0.05	ϕ_2 22.20 ± 0.03	*L* 12.50 ± 0.10
12.12 .2007	8.10 AM	24.40	22.10	12.35
	9 „	24.35	22.05	12.35
	10 „	24.35	22.05	12.30
	11 „	24.40	22.10	12.35
	12 Midday	24.45	22.15	12.40
	1 PM	24.40	22.10	12.40
	2 „	24.35	22.05	12.35
	3 „	24.35	22.05	12.30
	4 „	24.35	22.05	12.30
	5 PM	24.40	22.10	12.40
	6 „	24.35	22.05	12.40
	7 „	24.40	22.10	12.40
	8 „	24.45	22.15	12.40
	9 „	24.45	22.15	12.35
	10 „	24.40	22.10	12.35
	11 „	24.35	22.05	12.30
	12 Midnight	24.35	22.05	12.30
	1 AM	24.35	22.05	12.35
	2 „	24.35	22.05	12.40
Min.	-	24.35	22.05	12.30
Max.	-	24.45	22.15	12.40
Avg.	-	24.40	22.10	12.35

Readings in Table 6.3 are now shown in Graph 6.3.

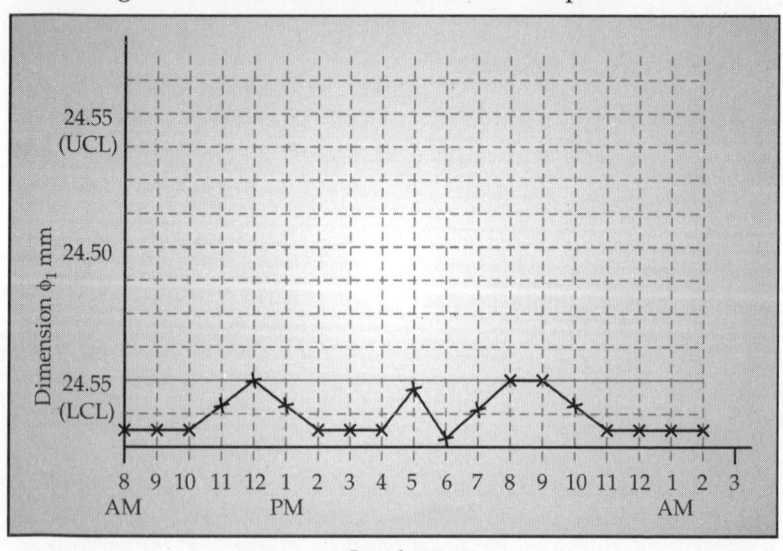

Graph 6.3

Data in three tables is shown as an example. In practice, sampling is done over a long period of time say, days, weeks, months or even year. Bulk of data helps in analysing the reasons of deviations. Data may reveal that deviations are erratic or having a definite pattern.

Analysis of Data

Referring to Table 6.1 and Graph 6.1, it can be seen that dimension ϕ_1 of component has crossed LCL (lower control limit) on two occasions. First crossing of LCL is from midday to afternoon and second crossing of LCL is around 7 to 8 PM. Probable reason for such a deviation may be that during midday, temperature of mould and injected plastic rises by a few degrees centigrade, enough to increase shrinkage of component after ejection from mould.

Crossing of LCL around 7 to 8 PM cannot be due to rise in mould and melt temperature due to ambient temperature as evenings are cool as compared to daytime. So there must be some other reason for deviation. It may be increase in supply voltage hence temperature of plasticizing cylinder increases. It may also be due to overenthusiastic operator 'playing' with temperature or mould stay setting.

Not losing sight on quality assurance, reasons of deviations should be examined. It may be checked if mould temperature control system is working properly and with reliability. If not, repair the system or install more effective and accurate mould and melt temperature control. Now look at other deviation around 7 and 8 PM. Practically it is occasionally found that over-enthusiastic operator slightly increases the speed of machine with the intention to produce more components in the shift. If the operator of injection moulding machine is not allowed to disturb the setting of various machine operational parameters, then it is indiscipline on the part of operator.

Fast running of machine may adversely affect extent of shrinkage of component after ejection.

On examining Table 6.2 and Graph 6.2, it is evident that some controls have become defective at 9 PM and continued till end of shift at 2 AM. On inspecting the machine it is found that plasticizing cylinder temperature control has become defective. Temperature of front zone is uncontrollably rising. Consequently, plastic melt is at higher temperature than specified and set. Since no attempt is made to compensate for the rise of melt temperature by reducing mould temperature and/or increasing cycle time, components coming out of mould are on higher temperature than normal. Consequently dimension ϕ_1 has crossed LCL.

Studies carried out over a period of month or two, can reveal if the defect detected is repetitive in nature or occurred once. If the defect(s) are repetitive then quality assurance policy calls for permanent and reliable repair of system or to replace components or units which are prone to function erratically.

On study of Table 6.3 and Graph 6.3 it is found that dimension is consistently below lower control limit (LCL). On examining, no fault could be found in the system which may cause consistent defect, that is one in all components are below LCL. On further investigation about supplies of polypropylene it is revealed that the grade is unapproved. It is not the one for which process parameters are already established and put in use. To avoid re-occurrence of such an error, more stringent procedure may be adopted for checking quality of plastic granules received. First thing is to read invoice and information printed on bags. All such actions would lead to quality assurance.

Throw away plastic syringes shown in Fig 6.2 (*b*) is another example where quality assurance is of utmost importance. It is not only the dimensions important, it should be sterilized properly so that no germs are left clinged to syringe. Syringes are then sealed packed. Quality assurance starts from factory/works designed for dust-free atmosphere. High degree of hygienic conditions are maintained. Syringe sterilization and packing should be fully automatic. At no stage syringes should be handled by human hands. If at any stage of production it becomes necessary to touch the syringes, it should be done after wearing suitable gloves which are sterilized. So all these arrangements are 'built-in' quality assurance system. Although testing of syringes for proper sterilization can be in test laboratory. It may be done by samples from packing line which are periodically tested. Suppose some samples are found to be contaminated then what can be done? It is not like machine which have controls for various parameters. Maybe production has to be stopped. Cleanliness and sterilization process to be examined for possible source of contamination.

Above example is given to emphasize that products of health care, treatment and intervention need 'built-in' quality assurance environment. Beverages and food products are another example where quality assurance can be attained by 'built-in' production processing system. In dairies, milk after pasteurisation should be automatically filled in bottles or pouches without being exposed to atmosphere and human touch to such an extent that there is no possibility of contamination.

Management of Quality Assurance

Any production system is prone to wear and tear, contamination. This may lead to deviation from quality norms. It is highly undesirable. Therefore, suitable actions or precautions are taken continuously on day-to-day basis. It may be termed as management of quality assurance which may consist of following:

- Preparation of quality manual
- Training of workers
- Display of instructions
- Collection of data
- Processing the data on computer having quality analysis software
- To take corrective measures as spelt out by computer and considered appropriate
- To manage that the whole of the work related to quality assurance is done correctly and regularly

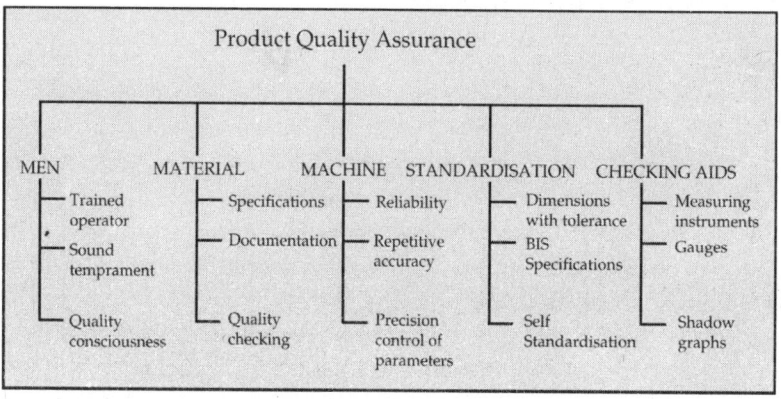

Preparation of Quality Manual

Production process may have a number of machinery or group of machines at different places in the plant. For example, fabrication, polishing, plating, vacuum metalising, assembly and packing. Quality manual may have chapters for each group or section of factory/production shop. Each chapter should contain drawing of component or assembly, description of its function individually or in combination with other components, dimensions with tolerances, input material specifications, finish, etc. Manual may also contain instructions regarding testing of components for strength, finish, internal stresses, light fastness, heat tolerance, effect of moisture, etc. whatever is applicable to a particular component.

Quality manual may include sample tables, graphs which should be filled by quality inspector on specified time or for each lot of specified numbers or quantities. Sample size may also be mentioned. Normally quality inspectors are not allowed to carry out any adjustments or corrective action. Quality inspector should inform production supervisor by showing him charts or graphs which are indicating quality deviations, in turn, production supervisor may or may not take any action immediately. There has to be a close coordination between production supervisor and inspector.

This is very important to clarify here that production supervisor is not expected to put forward the reason for defective quality production as quality assurance examiner did not inform him. It is the responsibility of production supervisor to be vigilant so that excellent quality norms are maintained.

Training of Operators and Workers

Training of operators of plant and machinery and other workers is essential for the management of quality assurance. After all, human beings are created to control the machines. Machines 'control' human beings only to such an extent for which those are made and programmed by human beings. Plant and machinery of any production process are operated in a definite pattern and sequence of activation. How correctly and efficiently an operator handles the machine or plant depends upon the abilities and knowledge of various controls and sequence of operation. Operator should be capable of understanding various sounds, motions, smells, heat, etc. normally produced in production process. Operator should be capable of noticing if anything abnormal happens. He should also know what quality aspects are acceptable and what are the deficiencies which may render components or assemblies rejected.

If worker or operator is placed to work or control a machine with quick and brief instructions, he may make mistakes, causing loss, damage to plant and machinery and even quality degradation in output from machines. Therefore, great emphasis may be laid on training of worker or operator before putting them on job. Training should be in a well-planned manner. Period of training and syllabus depend upon the complexities of production processes. Training programme for an assembly marker or packing worker may be simple and of short duration as compared to training programme for machine or plant operator.

Two typical training programmes are written in following paragraphs.

Training programme for an assembler or packer

Theory: ------------- 15% time ---------- 3.5 hours
Practical: ------------- 85% time ---------- 20.5 hours

Total 24 hours

Theory

- Material and design of packing material
- Differentiation between acceptable and unacceptable materials
- Quality aspects in packing
- Standard output of final packing

Practical

- Receiving packing material from store
- Inspecting packing material for quality
- Arranging packing material on table or hopper of machine or assembly line
- In presence of training supervisor, packer should carry out packing operation. At the same time listening to training supervisor, he may give guidance for carrying out packing operation

- To carry out packing operation without presence of trainer
- Trainer may call in quality inspector to check packing done by packer
- Packer should be attentive and listen carefully to quality inspector. He may point out some quality deficiency in packing

After training of 24 hours, operator may be put to do regular packing work.

Training programme for a machine operator

Theory: --------- 20% --------------- 9.6 hours

Practical:--------- 80% --------------- 38.4 hours

Total 48 hours

Theory
- Purpose of machine
- Brief working principle of machine
- Construction of machine
- What to check before starting machine or plant
- Safety precautions
- How to start machine
- How to carry on production
- Quality requirements of component or product
- To get familiarised with sounds and vibrations of machine and plant during proper functioning
- To get familiarised with smells so that if some unusual smell develops, operator should be able to notice

Practical: Practical training may be imparted in the presence of trainer. Operator may be asked to repeat start and stop of machine/plant many times. Trainer shall watch. He should correct the operator if he is going to make mistake. In this way operator would attain confidence.

Operator may now be left alone to run production. However, for few days, production manager or supervisor should keep an eye on quality of product for some time till good amount of confidence is attained that operator would handle machine and/or plant properly and would produce good quality product/component, taking care of safety precautions.

Display of Instructions

It is a useful shopfloor practice to prepare pictorial display in language which is generally understood by recruited operators and workers. Display may be a small board fixed near each machine, unit or production line. This helps operator to correct himself. Management may have an advantage that operator or worker cannot find excuses in case something goes wrong with quality norms due to operator not handling the machine as per displayed guidelines.

Collection of Data

Operator may be instructed to keep on collecting components so that about ten components are collected each hour. Quality inspector may keep on collecting sample components from machines and plant of all the sections. Collected samples would be measured and dimensions recorded in a suitably drawn formate. Attributes like finish, colour, etc. may be recorded in attribute nomenclature like poor, good, excellent for finish, for colour, shade correct or incorrect.

Collection of data may not be a permanent work. Once corrective measures are taken over a period of time and quality norms are stabilised, quality assurance may be given. Once this situation is achieved, frequency of collection may be altered according to need.

Processing the Data on Computer

Plant Manager, who is managing complete production process, might not be interested in going through all the data recorded by quality inspector. Most likely Plant Manager would like to know the following.

- Which particular machine is giving much variation in quality.
- Is maximum deviation in quality norms taking place when a particular operator is handling the machine/plant.
- What is the minimum, maximum and average deviation.
- Most probable reasons for deviation in quality.
- Suggested corrective measures.

Since loaded software in computer processes entered data, almost instantaneously plant manager or general manager can see the quality trends of ongoing production on their computer screens without leaving their tables.

Corrective Measures

On regularly having a look on quality trends and knowing the probable reasons, a general manager (GM) may order critical observation of a particular unit of a machine, which is most likely to be giving repeated trouble, hence adversely affecting quality trends. GM may decide to replace the unit with a most advanced unit which has a proven reliable performance, experienced by other users of unit. Since this type of corrective measures involve expenditure of money, GM may have the authority to go for upgraded unit. This does not mean that Production Manager cannot contribute to the improvement. He may suggest and recommend for replacing existing unit.

In case, on investigation it is found that raw materials like plastic granules, aluminium sheets, brass sheets or zinc calots are not up to the mark in quality, do not have qualified quality then corrective measure would be to obtain replacement from supplier. If supplies of raw materials or input components are not up to the mark then it might be necessary to select alternative supplier.

There is a possibility that variation in quality of components or product is due to human error on the part of particular operator or worker. In spite of all automatic controls, there are occasions when operator has to make some adjustments or to stop machine if output is not of qualified quality.

Components or product having qualified quality, from producer's point of view, may give some performance trouble while in use at customer's place. Simple example may be of precision bolts with hexagonal heads used in assembly of a machine or equipment parts. Sometimes location of bolt in machine is such that for screwing in the bolt takes long time as spanner, open end or ring, can rotate bolt by a few degrees only. This may cause irritation to the fitter. In due course of time this difficulty is brought to the notice of bolts producer. Although bolts are of standard specified quality, then what corrective measure can producer take? He may introduce some suitable quantities of some bolts with a slot for screwdriver. It may be called a 'corrective measure'. It will be possible for fitter to use a screwdriver to screw in the bolt quickly in machine housing and then finally to tighten the bolt with suitable spanner. This means that an additional operation of slot making on the head of bolt would have to be introduced. It means procurement and installation of slotting machine. Is the investment justified or not?, can be decided by a top authority of organization who is looking after marketing, finance and complete operations of the particular bolt producing company.

Regularity in Quality Assurance Related Work

This is an administrative aspect. In a factory or works there may be a number of sections of production or assembly operations. There may be a few inspectors of quality checking to take care of all products. Sometimes one or two persons do not turn up for work. In such an event, production manager might make alternative arrangement. He may place some other person from office work (say, a data processor) to do the job of quality inspector till absent person joins duty.

7

Automation in Production Processes

Purpose of Automation

Any production process may be expected to give product of cent per cent qualified quality. Qualified quality may mean specified quality, quantity norms, finish, get up, packing and even loading for shipment. Production of consumer items, durables or any mechanical, electrical, electronic, optical or acoustic product may have passed through various stages of production. Operations in stages may be putting of input items or material in place, mechanical operations, heating or cooling, weighing, time dependent operations, finishing, unit packing, carton packing, counting, marking, etc. Most of the operations, if carried out by human beings, would be prone to variations because of variations in human senses of vision, touch, hearing, smell, reflex actions, physical ability, etc. Most probably these human senses vary from person to person. Even in an individual, senses may vary from time to time due to multiple and complex reasons and circumstances. Many a times an individual may have sharp senses but does not take work seriously. Consequently, his sharp human senses do not produce good results. There are many individuals who have some physical or mental impairment but carry out good work while on job. This may be due to the fact that they have a superior ability to concentrate.

Just think of matchbox of about half a century back. Labels were found unsymmetrically pasted, matchsticks were used to be sometimes less or more in number, fitting of tray in box was too tight or loose, printing quality of label was not attractive, amount of burning material on the stick was varying so much that it could be seen. It may also be visualized that rejection must have been quite high. It was all due to the fact that almost all the operations of production of matchbox were manual. With passing time, overall developments in materials, technologies, marketing strategies and tough competitions compelled producers to produce better quality product with minimum production cost. This could be gradually made possible by replacing human activities by ingenious mechanisms like pick and place, orientation, etc. It may be called elements of automation.

Purpose of automation is to limit variations in process, stability in quality, minimize rejection and reducing cost of production.

Elements of Automation

Following are commonly used elements of automation.

- Hoppers
- Feeders
- Orientators
- Pick and place mechanism
- Variety of mechanical motion mechanisms
- Sensors
- Electrical and electronic gadgets
- Microprocessor
- Actuators
- Varieties of mechanisms
- Computerized controls
- Robotics

Hoppers

Hoppers are those parts of machine or plant in which raw materials for a production are filled. Raw materials may be in the form of powder, granules, paste, liquid, balls, tablets, capsules, grains and even small components. None of raw materials is in an orderly manner. They are just poured in. There are large number of varieties of hoppers. Their design and size depend upon the type of raw material to be filled, quantity to be retained by hopper at a time, this means its capacity.

Hoppers in which coal is unloaded from railway wagon are huge and sturdy. In cement producing plants, stones are dropped in hoppers which are heavy, strong and sturdy.

In hoppers of injection moulding machine or extruders only a few kg of plastic granules are dropped. These are small sized hoppers in which about 25 to 50 kg of plastic granules can be dropped. In dairy plant, milk is poured by vendors from their containers to receiving hopper of plant. In pharmaceutical production plants, there are hoppers fitted to various machines. These hoppers are filled with medicinal powder for producing tablets. These are only a few examples which are mentioned here. There may be a large number of such examples.

Feeders

In almost all production processes raw materials or components for further operations are fed in an orderly manner. Quantity of feeding at a time is fixed and synchronized with the movement of other mechanisms. When it comes to feeding of components, they are to be fed in a particular position. Take example of a toothpaste tube cap. In a placement and screwing machine, cap should always be fed unidirectional, either having threaded side down or up. Therefore, design and construction of feeder are done in such a way that it orientates components in a specific position before finally feeding the component to other unit or station of machine.

There are metering feeders also. These are used to feed quantities of granules, powder of specific weight for each stroke of machine. There are a large number of varieties and designs of feeders. Some feeders are integral part of a machine. Some are separate units attached to machine. Such type of feeders are synchronized with the machine by mechanical or electrical means. Feeders are normally fed by small quantities of raw material or components from hopper automatically. In some cases this feeding may be manual. Feeders are equipped with a system which stop entry of raw material or component from hopper if feeder becomes completely filled up. Generally speed of filling of feeder from hopper is higher than the output from feeder. In this way feeder is always remained filled up. Hence no stroke of machine goes blank.

Orientators

Many components are similar from both the sides like a simple plain washer or a cylindrical spacer of metal or plastics. These types of components can be fed from any of the two directions. There is no need for orientation of one or the other side before feeding to machine head. But still there is a need for orientation of cylindrical spacer, for example, in vertical or horizontal position. If the shape of both the ends of component is different then it is necessary that component should be fed in a particular direction. For example, a sheet metal cup is to be fed in a trimming machine with open end of cup always in the direction of lower trimming roller. This is achieved by some specialized construction of chute, channel or mechanism. Whole process of bringing the component in particular direction is called orientation. Orientation mechanism may be from simple to very complicated design and construction.

Pick and Place Mechanism

'Pick and place' are those mechanisms by which component or raw material is picked up from one place and placed on another location. There are a large number of varieties of pick and place mechanisms. Mechanisms may be heavy duty or very delicate to handle tiny components. Lifting of heavy steel plates from a platform, carrying it onto the bed of radial drilling machine and placing it between the guides is an example of heavy duty 'pick and place' arrangement. On the other hand, a delicate pick and place mechanism vacuum lifts electronic chip and places it on a specialized location on PCB (printed circuit board).

Pick and place mechanism may have an arm which can axially move in all the three X, Y, Z axes. Moreover, picking arm may rotate in any direction. It is not necessary that all the movements and flexibilities are available in all mechanisms. It all depends upon as to what actions are to be performed by pick and place mechanism.

In most of the cases speed of any motion is very important. It has to be synchronized with the speed of machine. Pick and place cycle time may range from say, five minutes for heavy work like transfer of steel sheet to two seconds for transfer of tiny electronic chip. Many a times accuracy of placement of a component is very important together with high movement speed. For such jobs, light, strong and with minimum possible vibration mechanisms are designed and built. As already described, pick and place mechanisms are used where both picking and placing stations are fixed and at predetermined location. In case location of item or component, which is to be picked, is not defined, it may be anywhere in a specified area then a robot may be used as it would first identify as to where target item is located. It would travel to that place and even make sure if the item being picked up is desired item or some other similar item. After picking, robot would travel up to the location where item is to be placed. It would then accurately place it after ascertaining if placing location is vacant and OK.

Variety of Mechanical Movement Mechanisms

Various mechanical movements are required to perform different functions in innumerable machines and plants. For a given space, speed, force, torque, movement, etc. a number of machine elements are required. Size, shape, material of construction, properties, etc. of machine elements depend on design of function required. Hence there may be innumerable combinations, but basic working principle remains the same. It is, therefore, necessary to understand basics of machine elements, movements and mechanisms to appreciate any automation system. Following are some essential machine elements and mechanisms.

- Reciprocating movements
- Rotating motion
- Combination of above
- Planetary movements (motion)
- Movement in a geometrical pattern like paraboloid, epicycloid, etc.
- Leverage links
- Motion adoption by gear trains, rack and pinion
- Conversion of motion
- Combination of cams and followers
- Universal joints
- Pneumatic, hydraulic and electric actuators
- Precision stepping movements

A rod, block or spindle moving to and fro along its axis is called a reciprocating motion. It may vary in amplitude and speed according to requirement. Force available at the end of reciprocating element may or may not vary during its complete cycle of reciprocation. It completely depends upon the design of mechanism which is providing reciprocating movement to machine element, a rod for example.

Any machine element may have rotary motion. It is not necessary that rotating machine element has to be round in shape. A rod of rectangular section turning along its axis may be called a rotating rod. If the same rod is not turning along its axis, may not be called a rotating rod.

In a particular mechanism there may be a necessity that a rod reciprocates as well as rotate. Rotation of a reciprocating rod may be a continuous rotation, partial rotation in one direction and then rotating back to its initial position. Rotary motion can be intermittent also. This means that rod rotates for specific degrees, stops and again rotates some specific degrees. In this way it keeps on rotating. Such types of movements are required in indexing mechanisms. There may be functional need that a rod moves to one extreme end of reciprocation and stops there to rotate by 180 degrees and then to reciprocate back to its other extreme end of reciprocation.

If a wheel, gear or rod is rotating on its axis and as well as its axis is travelling along a circle then it is called a planetary motion (movement). It is shown in Fig. 7.1.

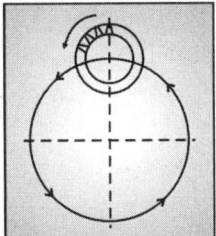

Fig. 7.1: Planetary Motion (Movement)

Planetary movements may be continuous, intermittent or backwards or forward for specific degrees.

Figure 7.2 shows a point at the end of a link. Construction of linkages should be such that shown point moves along the path shown by dotted lines.

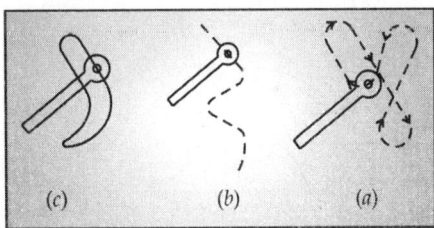

Fig. 7.2: Path of Motion

Leverage Links

Purpose of leverage links may be to transfer the movement from one mechanism to another as it is or to modify movement of driving element before transferring it to driven element. Levers are also used to reduce or magnify the length of travel of a machine element, say an axially driven rod. Links are also a via media to convert a rotary motion into linear motion and vice versa. There are instances where links are used to increase or reduce amount of force applied to a particular machine element. There may be examples of combination of two or more functions of links like modifying movement of a push rod together with applied force.

Motion Adoption

Motion adoption by gears, rack and pinion is very common in machines like gear boxes, gear trains in machine tools like engine lathe. Big and huge gears can also be seen in rolling mills, sugar mills, etc. Miniature gears made of metals and nylon are used in machines like meters, watches, toys, household gadgets, etc. Gears are also used in construction of automation devices. Purpose of gears is to transfer torque from one shaft to another. Set of gears are also used to create a ratio of rotation between driving and driven elements. A typical example of this is a reduction gear box where rpm of driving shaft is reduced to, say, one-third for driven or output shaft. In gear box worm and wormwheel are used. Worm and wormwheels are a kind of gears. In fact, there are quite a number of designs of gear boxes to cater to the needs of a mechanism. In automation mechanisms, use of gears is not uncommon. Gears are also used to convert rotary motion into linear motion. Example of this is a mechanical jack where a handle is rotated to lift the axil of a car.

Combination of Cams and Followers

Cam is a non-circular rotating portion of a shaft or a disc attached to it. Rotation of shaft may be continuous or for a certain degrees and then going back. Cam may be in the form of a disc with a profiled circumferential surface. Figure 7.3 shows a few cams with followers.

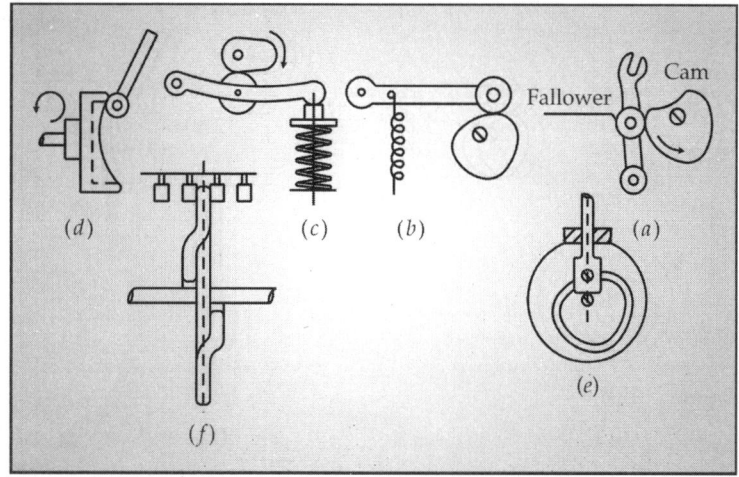

Fig. 7.3: Various Cams with Followers

Referring to Fig. 7.3 (*a*), this type of cam and follower can be seen fitted to sewing machines. With the help of this mechanism thread is wound on the bobbin. While the cam is rotating, follower lever keeps on moving to and fro, shifting the thread left and right for getting it wound systematically. This means that one round of thread is just touching the previous round. Once one layer is completed, second layer automatically starts getting wound. Figure 7.3 (*b*) is a round disc fitted eccentrically to the rotating shaft. Follower roller is touching circumferential surface of disc. Since the disc is fitted eccentrically, its rotation moves the follower up and down, down by spring tension and up by cam. Figure 7.3 (*c*) is another example of a conical looking cam. Rotation of cam keeps on pushing and releasing rocker arm. In turn, rocker arm pushes valve stem down against compression spring. This arrangement can be seen in automobile engines. Figure 7.3 (*d*) is a disc cam, but in this cam, face is the working surface. Up to a certain depth from circumference face is given a definite shape that is thick and thin, thus forming a profile. Choice of such a cam depends on overall construction of drive system and space available for a particular movement of any machine element or unit of machine. Figure 7.3 (*f*) is a continuously rotating cam. It is a plate having curved section for certain degrees, say 50 degrees. It rotates between equidistant rollers of an under the table turret. As soon as curved portion passes between the rollers of turret, it gets turned softly, depending on the design of curve portion. Just before ending of curve portion, next gap between the rollers of turret is engaged. Therefore, for the rest of 310 degrees (360°–50°) turret remains stationary till curved portion arrives again. Figure 7.3 (*e*) is again a face cam. A round disc has a precisely machined channel in which roller of follower is engaged. With this cam axial movement of follower is positive for both to and fro movements. This type of cam may be found fitted in shaper, a machine tool.

Universal Joint

Universal joint is a combination of a few machine elements which make it suitable to transfer reciprocating or rotating motion from one shaft to another while axis of two shafts may keep on changing angle. This means that sometimes axes of both the shafts are in line and at other moment axes may be at an angle with each other. Universal joints are abundantly used in motor vehicles for power transmission from gear box to differential system of wheels. Figure 7.4 shows basic construction of a universal joint.

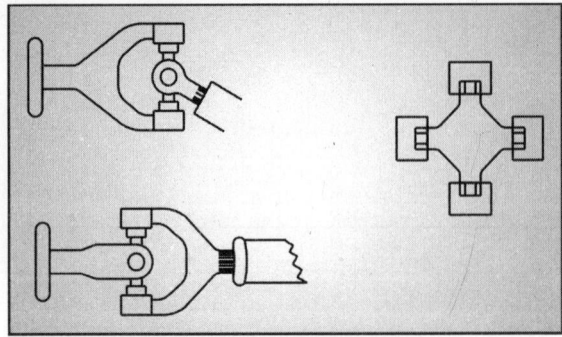

Fig. 7.4: Universal Joint

Universal joints are sometimes used in automation devices or arrangement.

Pneumatic, Hydraulic and Electric Actuators

Actuators are those devices which provide linear or rotary motion to other machine elements of machine. Actuators are basically pneumatic or hydraulic jacks. It consists of a cylinder, piston rod, inlet and outlet ports and seals. Figure 7.5 shows basic construction of a hydraulic jack.

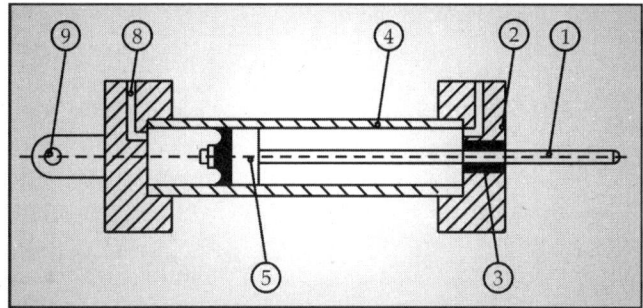

Fig. 7.5: Hydraulic Jack

① is piston rod which is connected to machine element or mechanism which is to be actuated. Cylinder ④ is fixed at a suitable place with the help of a flexible joint ⑨. Piston rod passes through cylinder head ② and rubber seals ③. Piston ⑤ is fitted with a pressure rubber seal ring. Fitting of piston and rubber seal ring is such that hydraulic oil cannot pass through running clearance between cylinder and rubber seal ring. When hydraulic oil pressure is applied through port ⑧, force generated due to oil pressure pushes the piston towards head end of cylinder together with piston rod. Flow of oil from pump to cylinder and cylinder to drain is controlled by a directional control valve in hydraulic circuitry. At a given moment, if oil enters a port then other port is connected to drain for oil to be pushed out by the face of piston. Hydraulic motors can also be called actuators because rotary and linear motions are provided by such motors.

Pneumatic jacks are almost similar to hydraulic jacks, with the only basic difference that linear movement cannot be precise as air is compressible. Hydraulic jacks can be operated precisely for a desired length of stroke. Pneumatics are used where no much force is required and travel of actuator can be restricted by a suitable stopper. Another factor in favour of pneumatic actuator is that the place is free of oil wetness.

Electrical actuators are generally AC and DC solenoids. These are normally required to actuate valves, pushers, selector flaps and opening of some kind of door in the outlet channel of a machine. Electric motors may also be rotary actuators, providing continuous or intermittent rotary motion. Stepper motors are devices to provide 'stepless' or 'with steps' circular motion of precise requirement.

Precision Stepping Movement

In this paragraph a typical example is described where precision stepping movement is an indispensable requirement. In extrusion of plastic film variation in thickness is controlled in a very close tolerance, say 0.01 to 0.03 mm. In sheet extrusion line thickness checking sensors are installed which continuously monitor the thickness of sheet. Signals from sensors are processed by microprocessors of attached computer. In turn, computer sends command to a stepper motor to rotate by, say exactly 11 degrees. The next moment command may be for rotation by 8 degrees in reverse direction. This rotation of stepper motor actuates servo system which actuates adjustment mechanism of extrusion die head to reduce or increase gap between die and plate to affect correction in the thickness of sheet. Another example of precision stepping movement is electric discharge machining (EMD). In spark erosion machine (or EMD) cavity in steel block is formed by means of a copper, carbon or steel electrode. Gap between the surface of cavity and electrode surface is very important and has to be maintained for a specific distance, say 0.09 mm. This gap for proper electrical discharge bombardment is critical. This gap is electronically sensed and translated into required up or down movement of electrode. This is a continuous process and is done by a stepper motor which drives a fine threaded shaft. Nut attached to EMD head or slide, converts circular motion of screw to linear motion of head up or down according to requirement for maintaining a definite amount of gap.

Computerised Controls

In almost all production processes in small, medium and large industries all types of controls are handled by computers which are loaded with suitable software. In case of adjustment of various process parameters which are interconnected are quickly controlled by computer. Manual control may not be possible because manual analysis of situation may take long time than its implementation. Computer does the job in a very short time. Installed software is such that appropriate commands are given to various controls.

Robotics

Just visualize a situation where a vendor places a bottle of milk at the doorstep early morning. There may be ways to bring the bottle to kitchen. One, get up, leave bed, go to door, pick up the bottle and bring it and place on kitchen table. Second option, employ a domestic help to do the job and pay handsomely for early morning engagement. Thirdly, install a conveyor belt from door to kitchen with control switches. Well, first and second options may be practical, but the third option is not impossible but absurd.

In a house, garage, office or production place there may be many small work which may be automated but would not be justified and almost impractical. Robot is a machine which is programmed to identify objects, find its way, handle the object as it should be or to replace it to another location. Robot is a self-contained machine which may be designed to copy human movements for performing a job. It needs little space for moving around shopfloor, in office, at home or even roads. There is 'near true' movie showing robot looking like a human being running on the road with a lady hand bag in hand. A security officer on a round takes that robot as a thief knocks the robot down just in front of a lady to whom robot hands over bag just before being knocked down. Security officer asks the lady if the bag is hers. Lady says, 'obviously it is mine, have a breathing disorder. He has just brought my bag which contains emergency medicine. He is just in time'. Robots are used in assembly lines of automobiles for carrying out spot welding and other operations. There may be many other situations where use of expensive machine, robot is justified. Another justifiable example may be use of robot by a bomb disposal squad. Robots are quite often used in research and development activities.

Examples of Automation

Automation is combination of a number of machine elements, mechanisms, control systems and source of driving power. Since automatic operation may be a combination of activities, some type of arrangement is required to synchronise the activities. This synchronization may be purely by mechanical means or through combination of mechanical, electrical, hydraulic, pneumatic and electronic. In modern automation system use of microprocessor, computer, etc. is common. Automation may be categorised in following groups.

- Receive and place
- Pick and place
- Static
- Performer

In 'receive and place' category of automation, components, material or input enter the automation device by itself. Automation device only carries it or just pushes the component to its due position. This means that item or object is placed on its correct position.

'Pick and place' is that automation where one of the functions of system is also to go to pick the item and then place it at the desired position.

'Static automation' is that arrangement where there is no moving part by virtue of any kind of drive system. 'Performer' automation does not necessarily handle any component or product but a number of operations are performed in a synchronized manner. This type of automation is abundantly found in automatic plants and machinery.

Author is providing herewith brief description of a number of automation systems. Care is taken to cover a variety of combinations of movements of machine parts. Descriptions are supported by suitable typical figures.

Receive and Place Type

Figure 7.6 (*a*) shows a plastic component of about 28 mm diameter and having a neck of 8 mm diameter and 3.5 mm height. Crown diameter is 23 mm and height is 2.8 mm. Weight of component is about 2.4 grams. These components are to be fed to a cylindrical component which has a coaxial rod of about 3.8 mm diameter. These components came at feeding station to be loaded on a turret having equidistant seats. During stay period of turret, plastic component should get in the component having a coaxial rod. Plastic components are filled in hopper ①. It has a rotating disc which is attached to a shaft ③ which constantly rotate with a speed of about 15 rounds per minute. Rotating disc has a shallow conical shape having many shallow slots of about 29 mm width. While disc ② is rotating, a number of components get into shallow slots of rotating disc. Components coming in front of feed channel get slipped into feed channel ④. In this way channel remains filled up. At the outlet end of channel there is a rocker arm with stopper pins. Dimensions of rocker and pins are so maintained that either of pins restrict movement of plastic component. As soon as a seat of turret with component reaches feeding station, rocker arm ⑦ releases one component while not allowing another component to slide down. Just before turret starts indexing, rocker arm pin restricts movement of plastic component. Rocker arm releases it only when next seat with component reaches in front of feeding channel. This process keeps on repeating itself.

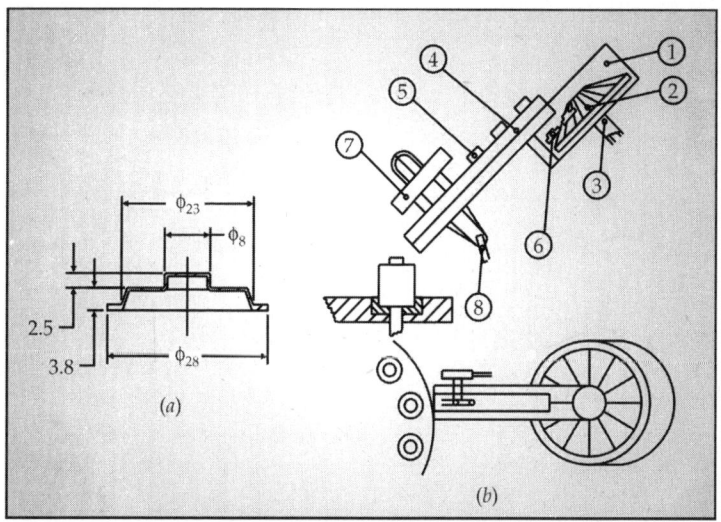

Fig. 7.6: Receive and Place System

Figure 7.7 shows another 'receive and place' mechanism.

Referring to Fig. 7.7, ① is magazine which is filled by shells manually or by a conveyor belt. ④ is an arm of machine which swings left and right by certain degrees, say 30 degrees. On the top of arm plateform, a small pick and carry mechanism is attached. It has a long tong ⑤ to prevent the shell falling down when arm moves to extreme left position. Pick and carry attachment is fitted with a holding arm ⑥. One end of holding arm has curvature to match shells to be held. Other end has a roller ⑧. Holding arm is pulled down by a tension spring ⑩. There is also a pin ⑨ which keeps the holding at such a position that gap (g) is less than diameter of shell and having a gripping force due to tension of spring. As soon as machine arm ④ swings to its extreme right position, roller ⑧ rides over a cam ⑦. Consequently, gap (*g*) increases and one shell gets dropped in the gap (*g*). As soon as machine arm ④ starts moving towards its left side position, roller ⑧ gets released from cam ⑦. Hence full force of spring is transferred to ⑥ for firm grip of shell in the gap. When machine arm reaches its extreme left position, it stays there for a while. During this time a pusher comes to push the shell on to machine spindle. After a little delay, machine arm swings back again to right to receive another shell from magazine. In this way shells are automatically received by mechanism and placed to such a position that other machine element or mechanism 'takes charge' of shell. In this way process keeps on going.

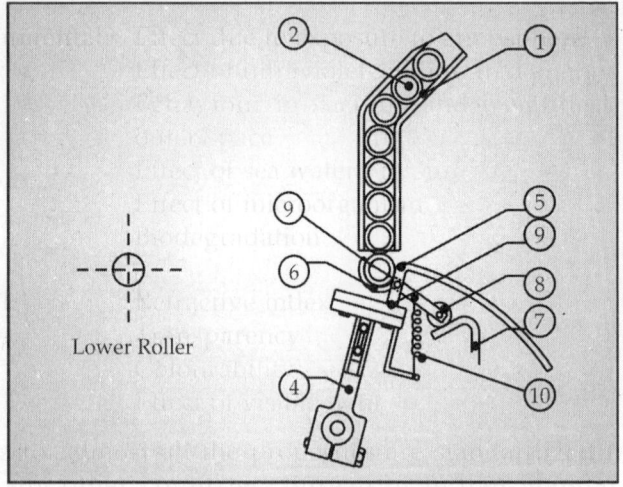

Fig. 7.7: Another Receive and Place System

A third example of 'receive and place' mechanism is shown in Fig. 7.8.

It is the mechanism which places the calot in front of extrusion die ⑦. ① are calots which are guided in a feeding channel ②, ③. Feeder ④ is at such an extreme left position that a calot fells between its sliding unit and spring loaded tong ⑧. In this position tong ⑧ is in slightly open position against spring tension. As soon as feeder ④ moves towards right, tong grips the calot firmly enough that it does not fall. Feeder ④ moves right to reach its precise position. In this position centre of calot coincides with the centre of extrusion die and punch. Punch comes with high velocity and takes calot inside the die. While feeder is in its right side position, no calot can fall as lowest calot is held by upper surface of feeder slide. All the movements are mechanically synchronized.

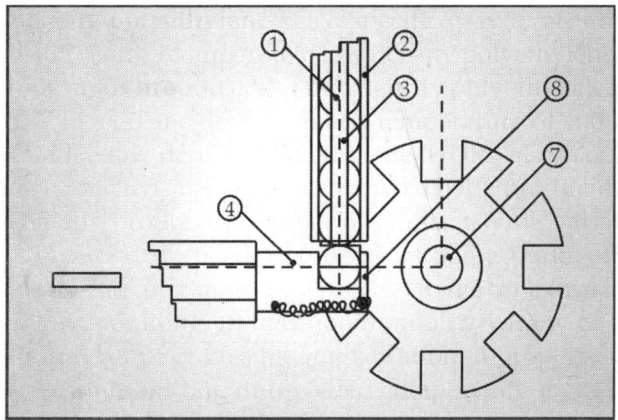

Fig. 7.8: Third Receive and Place System

Pick and Place

There are many situations in production processes where automation is done for picking up the component and placing it at a particular position. In this situation, component does not come in for carrying and placing. Component has to be picked up.

Few examples of 'pick and place' automation are briefly explained with the help of Fig. 7.9.

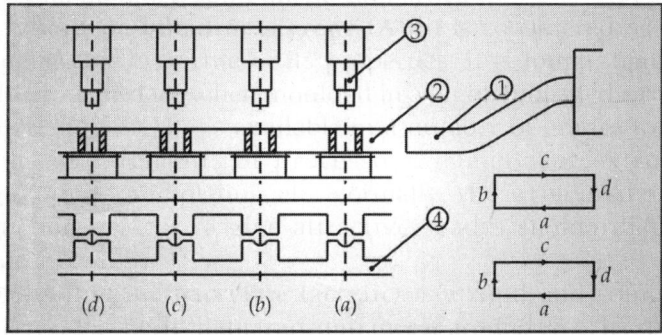

Fig. 7.9: A Pick and Place System

Figure 7.9 shows a power press which has a series of dies and punches. Input to this press are shells of about 20 mm in diameter and 13 mm high. These shells are required to be converted to shells of 10 mm diameter. This is achieved by progressive draws in four stages. ① is a channel which connects a feeder to station Fig. 7.9 (a) of press that is first draw die and punch. On next stroke of press, a shell present on die Fig. 7.9 (a) is drawn. Now it is to be transferred to position Fig. 7.9 (b). On second stroke, Fig. 7.9 (b) is to be transferred to Fig. 7.9 (c), and another shell duly drawn from station Fig. 7.9 (a) to Fig. 7.9 (b). In third stroke, Fig. 7.9 (a) is to be transferred to Fig. 7.9 (b), 'b' to be transferred to Fig. 7.9 (c), and 'c' is to be transferred to Fig. 7.9 (d). This is achieved by pair of transfer arms (in strip form). Motion of arms is mechanically synchronized with the movement of power press mini rams which are operated by a heavy and sturdy cam shaft. Movement of pair of arms is indicated by a movement depicting sketch. Referring to this sketch, arms grip all the components on four stations duly ejected by ejector rods from below the dies. Arms move one pitch of dies. That means central distance of one die from next die. Arms open out leaving the components on dies and dropping the last one on a ejection channel. Arms then travel back to original position and close in to grip next set of components and repeating pick and place operation.

Next example is of a case where component is picked up from one turret of a machine and placed under the head of another turret of same machine. Figure 7.10 shows basic arrangement of two turrets and transfer arm.

Figure 7.10 (a) is that of a component which is mounted on the head of turret ①. Each head of turret ①, when stays at station s_1,

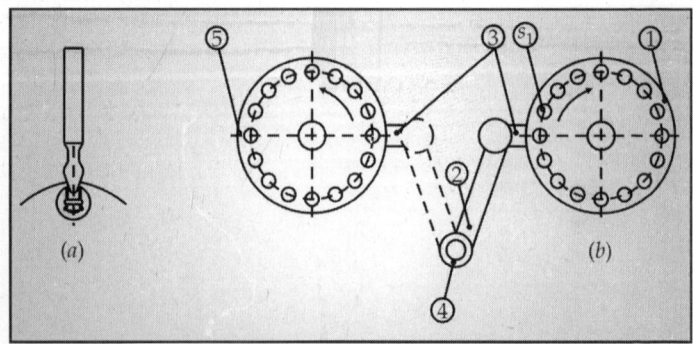

(a)   (b)

Fig. 7.10: Another Pick and Place System

has a component mounted. There is a transfer arm ④ fixed on the base of machine between two turrets. Vertical spindle of arm head ④ partially rotate clockwise and anticlockwise with certain angle. On the top of spindle an arm ② is attached. At the other end of arm is fitted a gripper ③. While the turrets are in rotating mode, arm is also moving. When turret stops, just at that moment gripper ③ reaches and grips the glass stem of the component 'a'. Immediately after gripping, arm swings towards turret ⑤. While arm is swinging, gripper ③ also swings with component towards head of turret ⑤. It reaches turret ⑤ just after it stops moving. Now component is just over a pusher which pushes glass stem (tube) in the head of turret's over head ring carrying heads with rubber grippers. Design of arm, its gripper and movement is quite complicated and unique. It is, therefore, explained in detail with the help of Fig. 7.11.

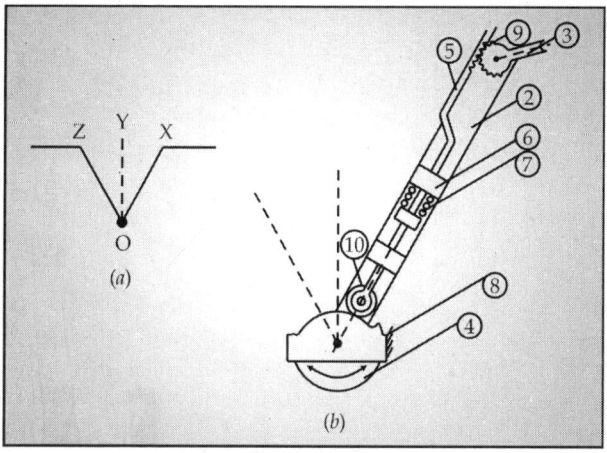

Fig. 7.11: Gripper Movement

Figure 7.11 consists of two drawings 'a' and 'b'. 'a' is a line diagram depicting position of transfer arm ② and gripper ③. OX is the position of arm and gripper when it has picked up component from turret ①. Arm then swings in anticlockwise direction. Linkages between arm and gripper are so designed and made that gripper also starts swinging in anticlockwise direction with its swinging point at the end of arm OX, when arm reaches position OY, gripper has already swung to position almost parallel to arm.

When arm reaches at position OZ, gripper has also swung further in anticlockwise direction so that component is just above pusher rod of head of turret ⑤. Figure 7.11 (b) shows the mechanism which provides motion to gripper while arm ② swings. Gripper ③ has a rotating pin joint with a gear segment ⑨. A rack ⑤ is connected to gear segment. Rack ⑤ is rigidly fitted on an extension rod passing through holes in extended blocks ⑥, on swinging arm ②. At the end of extension rod there is a roller ⑩ which is touching a fixed cam ⑧. This cam is so fixed with machine body that it is independent of arm head ④ movement. Consequently, roller ⑩ of extension rod acts as follower of cam and operates gripper. All construction elements of this mechanism are so adjusted that gripper ③ is pointing towards right when arm is towards extreme right. When arm is at its extreme left position, gripper is pointing towards left. Since the arm ② is mechanically synchronized with movements of turrets, automatic picking of component (glass stem) by fingers (spring loaded or having vacuum system) and swinging to other turret keep on repeating itself with precision.

Another example of automation of 'pick and place' nature is where a glass tube is first oriented while coming down from a bulk hopper. Once oriented in vertical position a vacuum picker holds vertical glass tube, carries it to gripper of turret head. Basic construction is shown in Fig. 7.12.

A very interesting 'pick and place' automatic system is described here with the help of three figures. Figure 7.13 shows as to what is required (a) is a sintered glass bead. Approximate dimensions are

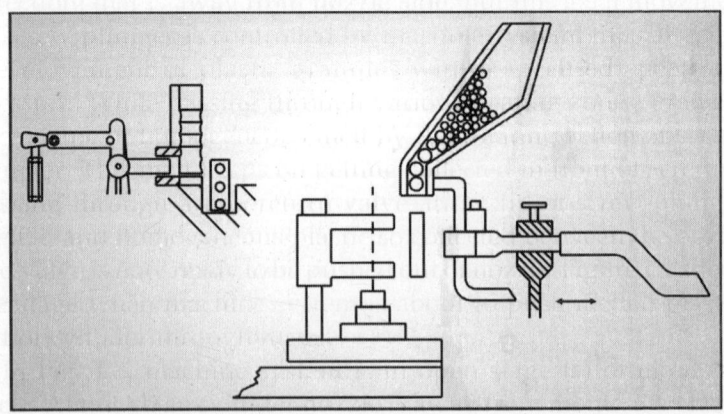

Fig. 7.12: Vertical Pick and Place Mechanism

given. (*b*) is the head of of a turret. Fingers ① are gripping two electrode wires. Some length of wire is above the finger ① and some below the fingers. Indexing speed is fifteen indexings per minute. This means that each indexing takes four seconds. This time of four seconds is total time of stay of turret and then getting indexed. Out of four seconds, stay time is about three seconds and one second for indexing movement of turret. Stay time is the period in which sintered glass bead is to be fed. Automatic pick and place system is such that glass bead gets dropped around upper electrodes (protruding wires) as soon as turret stops. System to pick such a small component and then to feed it around electrode is explained with the help of Figs. 7.13 to 7.15.

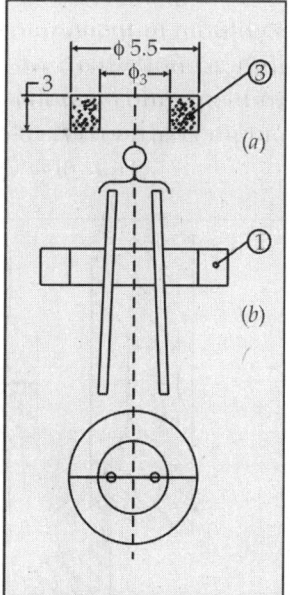

Fig. 7.13: Electrode and Bead

'Pick and place' system for feeding sintered glass bead consists of two sub-systems. First is shown in Fig. 7.14.

Referring to Fig. 7.14, ① is a bowl in which beads are filled to a certain level. ② is a tube which has an opening ⑭ at the top. Bore of opening is so much that diameter of bead passes through opening. Gap between opening ⑭ and feeding pin ③ is such that any bead falling on the side of pin does not pass through the gap and actually falls down in the bowl due to taper shape of tube at the top. ④ is a spindle which passes through bowl walls. On the spindle is mounted a strip ⑤. At the end of this strip are mounted picking needles ⑥. Spindle ④ rotates in a peculiar manner. While rotating intermittently, carry pins pass through 'mass' of beads. Level of beads is so maintained that only pin ⑥ dips into beads and not strip ⑤. When spindle ④ rotates, needle is filled with beads also move to such an extent that end of feeding pin reaches near vertical feeding pin ③, only a few mm away from vertical pin ③ top end. All beads ⑦ slide down. Some fall around pin ③ and some in bowl. Spindle further rotate only after axially shifting to side so that pin ⑥ is not in striking range of vertical feeding pin ③. Further rotation of spindle ④ brings the other pin ⑥ in position. In this way both pins ⑥ keep on feeding beads around vertical feeding pin. ⑫ is a ratchet operated by a cam underneath the table of machine where various cams are fitted. This ratchet rotates indexing wheel by 90 degrees. There is a cam ⑨ on the diameter of which cam channel is machined. This cam is permanently fitted to bowl's outer wall. There is a follower bracket with roller engaged in track of cam ⑨. When spindle rotates, axial movement also takes place due to hump in cam track. Design and dimensions are so maintained that following action takes place in sequence.

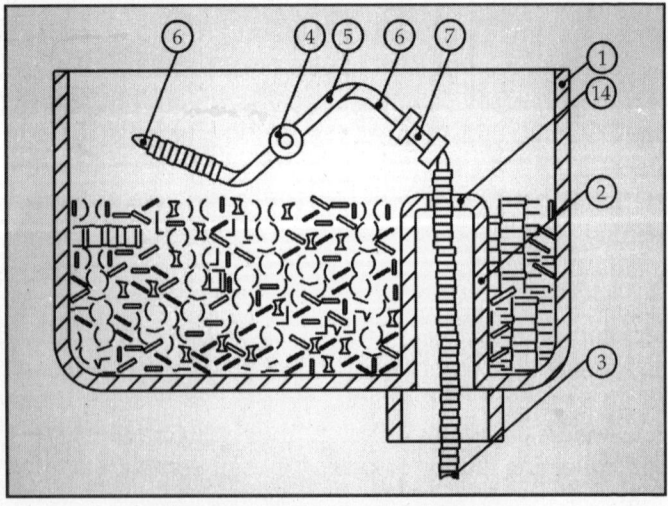

Fig. 7.14: Pick and Place Example

- Pick up pin ⑥ enters beads in the bowl
- Spindle rotates further till pick up pin's pointed end reaches near pointed end of vertical pin
- Spindle starts to rotate further together with immediate axial shifting
- Next pick up pin passes through beads, thus picking up few beads.

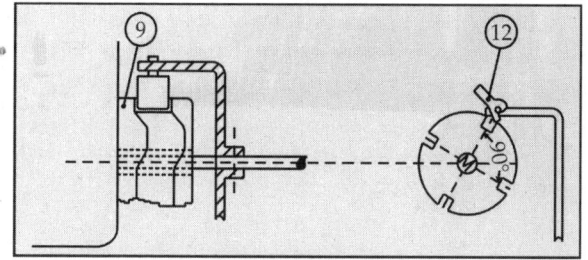

Fig. 7.14 (a)

Referring to Fig. 7.14 (a), one time stroke of ratchet during one complete indexing cycle of turret takes place. Each time spindle ④ rotates by 90 degrees, hence moving ratchet wheel by 90 degrees. This gives a little time to placing needle to stay aligned with vertical feed pin between two 90 degrees timing of cam. Now comes the description of system which holds the vertical pin in place and yet places beads around electrodes. This is explained with the help of Fig. 7.15.

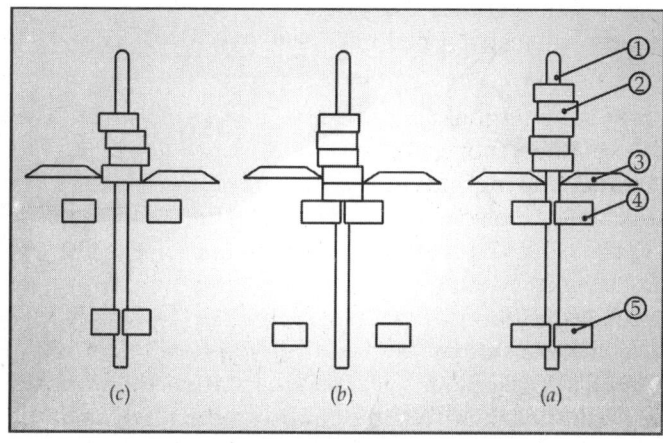

Fig. 7.15: Holding of Vertical Feed Pin

① is vertical floating pin. Pin is called to be floating as it is not fixed anywhere. It is held in place vertically either by finger pair ⑤ or ④. These two fingers open and close alternatively. When finger ⑤ holds the pin, finger ④ opens. Then after a laps of few moments finger ④ closes and finger ⑤ opens. Sharp edge finger pair ③ is meant to hold the beads from dropping and releasing only at a pre-adjusted sequence. It is worth noting that all the fingers ③, ④ and ⑤ get operated by cams underneath the machine table. All fingers have some mechanism for opening and closing. Mechanisms are connected to cam follower levers by means of connecting turn buckle rods. Sequence of operations is given below.

- All the three fingers are in closed position (see Fig. 7.15 (*a*))
- Finger ④ is holding vertical floating pin in place and fingers ③ and ⑤ get open (see Fig. 7.15 (*b*))
- Distance between lower face of finger ③ and upper face of finger ④ is a little more than one bead height. Hence one bead gets dropped and rests on upper face of finger ④
- Fingers ⑤ and ③ close and then finger ④ opens. The bead which was resting on finger ④ drops down to rest on face of finger ⑤
- Sequence repeats itself. Finger ⑤ opens. Bead gets dropped around electrode wires protruding above turret head fingers. Timing of sequence is so set that finger ⑤ opens only when turret has just stopped indexing. (see Fig. 7.15).

In following paragraph another typical mechanism, for the same work as explained above, is described with the help of Fig. 7.16 for picking and placing glass bead from feeding channel head to around protruded electrode in turret finger.

Referring to Fig. 7.16, ① is vibrating feeder in which beads are filled up to approximately specified height. Vibrator makes the beads travel continuously to feeding channel ② to keep it always filled up. Due to a stopper ③, beads do not flow out of channel. A swinging pair of gripper ④ is at picking station. At the position of swinging fulcrum there is a mechanism which opens and closes gripper and swings from picking station to placing station.

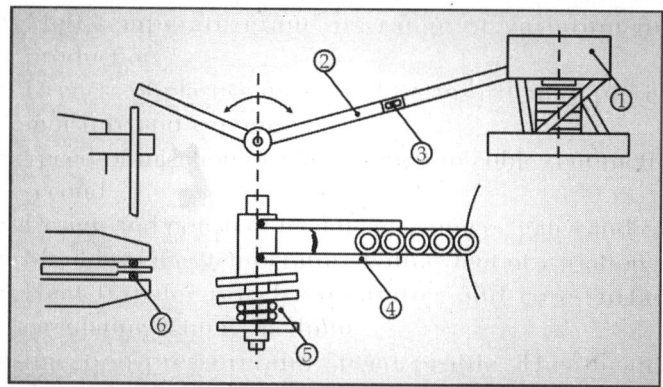

Fig. 7.16: Another Pick and Place Mechanism

Operating mechanism is synchronized with turret movement with the help of under the table cam, follower lever and turn buckle link. Sequence of operation would be as follows.

- Gripper pair reaches picking station
- Closes around a bead which is already there
- Swings to a position where held bead is just over and axially aligned with protruded electrodes on the top of turret finger
- Gripper opens up. Bead gets dropped around electrodes
- Gripper swings back again to picking station. In this way sequence keeps on repeating itself. Swinging gripper swings through a friction drive which rotates (reciprocative rotation) somewhat more than required swing. There are stop screws under the gripper to precisely adjust stop position on both picking and placing stations.

Performer

Performer is the name given to those mechanisms which do not directly handle inputs or outputs of any production process. There are mechanisms which ultimately operate particular unit with or without sensing presence of input component at a particular station. There may be innumerable examples of such mechanisms. In this chapter two examples are given and briefly explained with the help of figures.

First example is of a machine which is used to fill and mark a miniature container with precise quantity of paste.

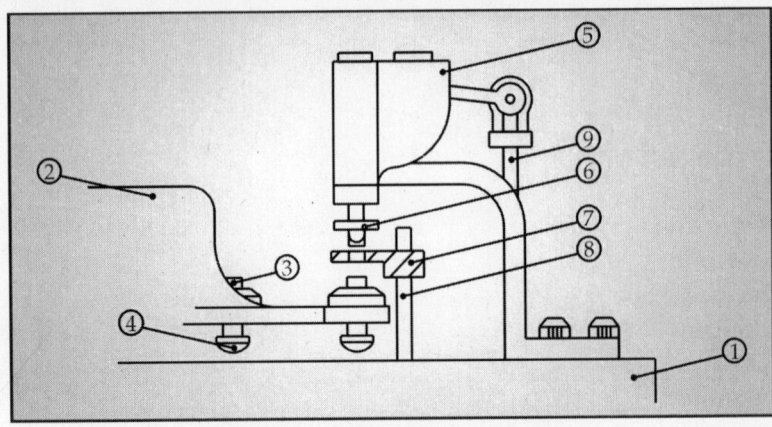

Fig. 7.17: Paste Feeding System

Following is the list of parts of machine.

Part No.	Description
1	Table of machine
2	Turret with eight or ten heads
3	Component placement heads
4	Ejection pusher
5	Paste dosing and feeding unit
6	Paste feeding nozzle
7	Sensor plate
8	Sensor plate rod
9	Paste unit actuating turn buckle rod

Basic construction of paste filling system of a typical paste filling machine is shown in Fig. 7.18. Machine has a turret which also moves up and down. It has a number of heads in which miniature components are fed. Turret gets indexed only when it is in down position. As soon as indexing stops, turret moves up and stays there for 'no indexing' period of indexing system. This is the position of turret when component is fed in turret head at component feeding station. At paste feeding station if component is present in turret head, lifts sensor plate which activates under the table mechanism to operate paste dosing and feeding mechanism. If by chance there is no component in turret head, sensor plate will not be lifted and hence mechanism under the table will not get operated. On third station automatic marking unit will mark the container (small component) by marking ink or stamping. Both being rotary units (not described). It needs to be briefly explained as to how sensor plate controls operation of dosing and feeding unit. It is explained with the help of Fig. 7.18. ⑨ is paste actuating push rod which has a follower tube ⑩ at the lower end. ⑪ is a cam which lifts follower tube to the extent of its throw. ⑧ is sensor plate rod which is inside a hole ⑮ of a plate ⑭. This plate is located on a pinion so that it can revolve. On the circumference of plate ⑭ there is a spring which keeps plate pulling against pin ⑮. There is a lever ㉒. Its one end is fitted to plate ⑭ and other having a follower roller ⑰. Lever ㉒ is actuated by cam ⑲ fitted on main cam shaft extension under the table. All the movements are synchronized with the movement of turret. Operational sequences are given below:

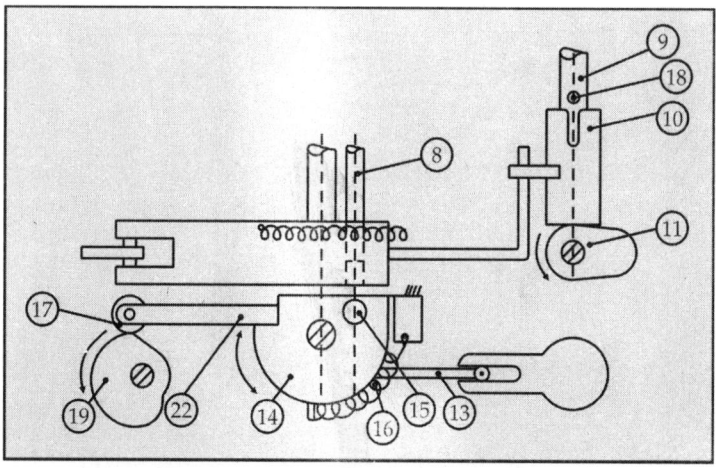

Fig. 7.18: Operating Mechanism

When there is no component inside turret head.

- Turret reaches at 'No Indexing' position and rises
- Since there is no component in the head, sensor plate is not lifted
- Rod ⑧ remains in hole of plate ⑭
- Cam ⑲ is at a position with 'no throw' just below follower roller. But follower roller does not move to touch cam
- Plate ⑭ remains stationary against pin ⑮

- Cam ⑪ actuates follower tube ⑩. This tube is below the pin fitted in pusher rod ⑨. No vertical movement of pusher rod takes place

  In case there is a component in the head then sequence of action of mechanism would be as follows:
- Sensor plate rod ⑧ would be out of hole of plate ⑭
- Less radius portion of cam ⑲ would be there and follower ⑰ with lever ㉒ will swing due to spring ⑯ tension
- Consequently lever ⑬ would rotate tube ⑩ to such an extent that its slot is no more under the pin ⑱
- Hence paste unit would operate and paste would be fed inside the mini container/shell

In this way machine would keep on working. Paste unit delivering metered quantity when mini container is present in the head and unit would not operate if there is no component in the head. There may be other type of construction of mechanism to achieve same function of the system.

Photographs

Photo 7.1

It is a line for a component on which number of operations are carried out. Machine visible in the front is fed with component. After the operation is carried out, component is automatically transferred to another machine by conveyors. In this way all the operations are carried out (*Courtesy: SISL*).

Photo 7.2

It is the end position of line, shown in photo 7.1 (*Courtesy: SISL*).

Index

❑❑❑